T0231931

From AI to Autonomous and Connected Vehicles

Digital Sciences Set

coordinated by
Abdelkhalak El Hami

Volume 2

From AI to Autonomous and Connected Vehicles

Advanced Driver-Assistance Systems (ADAS)

Edited by

Abdelaziz Bensrhair
Thierry Bapin

WILEY

First published 2021 in Great Britain and the United States by ISTE Ltd and John Wiley & Sons, Inc.

ISTE Ltd
27-37 St George's Road
London SW19 4EU
UK

www.iste.co.uk

John Wiley & Sons, Inc.
111 River Street
Hoboken, NJ 07030
USA

www.wiley.com

Library of Congress Control Number: 2021938230

British Library Cataloguing-in-Publication Data
A CIP record for this book is available from the British Library
ISBN 978-1-78630-727-9

Contents

**Chapter 2. Conventional Vision or Not: A Selection of
Low-level Algorithms**.   25
Fabien BONARDI, Samia BOUCHAFA, Hicham HADJ-ABDELKADER and
Désiré SIDIBÉ

**Chapter 3. Automated Driving, a Question of
Trajectory Planning**.   79
Olivier ORFILA, Dominique GRUYER and Rémi SAINCT

Chapter 4. From Virtual to Real, How to Prototype, Test, Evaluate and Validate ADAS for the Automated and Connected Vehicle? . . 125

Dominique GRUYER, Serge LAVERDURE, Jean-Sébastien BERTHY,
Philippe DESOUZA and Mokrane HADJ-BACHIR

Foreword 1

This book is the result of many collaborations between the engineering school, INSA Rouen Normandie, and the French cluster for automotive & mobility industry, NextMove.

The subject of the book concerns the field of driving automation (ADAS, advanced driver-assistance systems), in conjunction with the connected and autonomous vehicle (CAV) educational chair.

Created in June 2006 and located in the Normandy and Île-de-France regions (where 70% of French automotive R&D is carried out), NextMove embodies, animates and promotes the "Mobility Valley", an area of European excellence where solutions are invented, developed, tested and industrialized to meet the challenges of mobility today and tomorrow. Convinced that the future of mobility players in France depends on their ability to be competitive through innovation, NextMove weaves and animates a dynamic network between major manufacturers, SMEs, start-ups, higher education, research institutions and local authorities. NextMove brings together more than 600 member institutions and supports more than 500 labeled projects (including more than 250 funded projects). NextMove is the largest French network of scientific and technical excellence, bringing together more than 600 members from the French mobility and automotive industry.

It is within this ecosystem which encourages innovation that the idea was born to create an educational chair of excellence devoted to the study of the CAV, bringing together the SMEs of Groupement ADAS and INSA Rouen Normandie. At the origin of this initiative, there was a convergence of views between Gérard Yahiaoui SME VP of NextMove and University Professor Abdelaziz Benshrair, who agreed on the need to bring together high-tech SMEs and students of major engineering schools. Traditionally reserved for large companies, chairs previously seemed inaccessible to start-ups and SMEs due to the amount of funding. Thanks to the critical size of the Groupement ADAS and the common will of all the actors, this CAV chair was able to take shape.

Created at the initiative of NextMove, Groupement ADAS brings together several innovative SMEs in the field of driving assistance systems and autonomous vehicles. Its mission is to bring electronic and digital innovation to vehicles, to meet the challenges of road safety and driving comfort. Groupement ADAS provides embedded products, multidisciplinary expertise and interoperable open technologies. To continue to drive innovation, the recruitment of future talents is a key issue.

Indeed, the challenges to be met in the field of automated, autonomous and assisted mobility increasingly require the training of young engineers in new technologies. Innovation is not only the prerogative of major groups. Thanks to the creation of the "Connected & Autonomous Vehicle" educational chair, students will be able to choose this specialization during their fifth year studying computer science. For this, high-tech start-ups and SMEs will provide part of the 42 hours of courses proposed. Embedded perception, the modeling of the driver's behavior, the automation of driving and deep learning are all topics covered in this specialization. The possibility of being able to exchange daily with leaders of high-tech SMEs in the mobility sector will allow them to become aware of the formidable opportunities for expressing their talents, by tackling concrete cases during their internship.

Thierry BAPIN
NextMove, the French competitiveness cluster
for the automotive & mobility industry
May 2021

Foreword 2

"AI and ADAS", How Can Mobility be Improved?

The mobility context

Numerous statistical studies have been carried out over the last decade in France, Europe and around the world in order to quantify the number of road accidents and what causes them. At the European level (European Commission 2019), the problem is significant, with a total 1 million road accidents, 1.4 million injuries and 25,600 deaths per year. At a global scale, road accidents cause 1.35 million deaths and 20 to 50 million serious injuries or lead to a disability every year (World Health Organization 2019). In the case of France, there has been an average of 3,400 deaths per year for several years. But if we take into account the amount of people injured, the number rises to 70,000 casualties and injured persons. This represents a significant cost for society.

In addition, it clearly appears that the majority of accidents involve a human factor, including the driver (93%). These studies also show the predominance of a set of factors related to the driver's abilities and other factors deteriorate their ability to drive. These may be related to perception, interpretation, assessment, decision-making problems and, ultimately, problems related to their actions. More concretely, the driver may lack visibility, or – due to the complexity of road situations – have limited and/or superficial information, leading to a poor understanding or misinterpretation of the scene. In addition, there are also problems triggered by the deliberate violation of The Highway Code, alcoholism and drug abuse, lack of driving experience, and also fatigue, hypovigilance, monotony, or they may even just be distracted.

It is therefore essential to be able to help the driver in order to reduce the number of road accidents, as well as to minimize risk in road situations. This reduction can be applied by providing the driver with both informative (alerts, instructions, expertise) and active (automation, delegation and shared driving) assistance. These aids are intended to help, as much as possible, the driver's limitations regarding perception, decision-making, and their actions, without forgetting compensation for the driver's potential failures.

Perception and AI, endowing ADAS with sight and intelligence

These driving aids necessarily go through an essential and critical stage of perception. This type of perception must respond to a large number of constraints in order to guarantee a high level of performance, reliability and robustness. It must be dynamic and adaptive in order to be able to take into account the data coming from the embedded sensors, depending on their relevance, quality and availability. After being processed, this information enables the construction of cooperative local and global dynamic perception maps containing an estimate of the attributes of the "key components" involved in the road scene. In the representation of the road environment, these key components are made up of obstacles, the road, the ego-vehicle, the environment and finally, the driver. Having a knowledge of the state of all these key components makes it possible to have sufficiently comprehensive perception at our disposal, usable by a large number of embedded security applications in the short, medium and long term. However, active applications involved in the automation of the driving process require a truly high quality of service, ensuring the optimal reliability, confidence, accuracy and robustness of the attributes of road scene key components. For this, techniques based on the use of Artificial Intelligence (machine learning, deep learning, graph theory, expert systems, fuzzy logic, etc.) are being developed and applied for processing sensor data and enabling the production of detection, filtering, data restoration, tracking, identification and recognition functions. This processing is either applied per sensor or per sensor set (multi-sensor fusion). However, perception is often limited by the performance of embedded sensors and by the impact of adverse and degraded environmental conditions on the technologies used by the sensors. This is critical because a faulty or degraded sensor can jeopardize the proper operating of an active application like the ones used for automated driving systems. In fact, at levels 3 and 4 of the automation driving task, the driver is still present and at any time they can receive a request to intervene and to take over the driving task, when the system is no longer able to operate correctly. Unfortunately, it takes longer than 10 seconds (Merat 2014) for this transition (from virtual co-pilot mode to human pilot mode) to be achieved. This amount of time is necessary for an inattentive driver to perceive the current environment, to analyze and understand it, to make the best possible decision, and to act effectively on the vehicle's actuators (pedals, steering wheel, and potentially, the gearbox). Thus, to be warned of a disengagement from the automated piloting system, it is necessary to have information that is sufficiently distant in terms of space and time so as to make it possible to analyze, interpret and predict the future situation, and estimate its level of risk. The most effective way to extend this perception is to use the knowledge available coming from the other road users. For this, it is necessary to use means of communication (transmitters and receivers embedded in the vehicles or present on the infrastructure (road side)).

Communications, a requirement for developing far away perception and anticipating problems

These means of communication are called VANETs (Vehicular Ad-hoc NETwork) and follow the 802.11p standard. Different standards are available to manage these communication media dedicated to transport systems, for example, IEEE which proposes the WAVE architecture (Wireless Access in Vehicular Environments), ETSI which proposes ETSI TC-ITS architecture, and ISO offers the CALM architecture. In all cases, the goal is to provide media with a high Quality of Service (QoS). This QoS must guarantee the capacity of the communication medium to transmit data under the best possible conditions while respecting the criteria of availability, flow, transmission delays and minimum rate of packet loss (messages).

In the context of the development and deployment of automated vehicles, the use of communications is becoming a clearly essential and critical issue. Indeed, the information will be used to feed active applications (emergency braking, application of collision avoidance maneuver, vehicle platoon stability). In this context, the information transmitted by the means of communication will make it possible to update local dynamic perception maps, and also obtain extended dynamic perception maps (in range and in attributes). This extended perception makes it possible to feed the decision-making, path planning and action systems required for driving automation. The act of extending perception is also useful for anticipating and predicting future risky situations, as well as for generating optimal decisions (from a safety point of view) all the while controlling the maneuvers of automated vehicles. A delay, an interference, or a modification of the information contained in the data frame will have a significant impact on the extended perception of the environment and on data quality (inaccuracy, uncertainty, reliability, belief), making them potentially unusable under penalty of inducing a wrong and erroneous decision, and by extension, an accident. The same goes for the management of automated platoons. The only way to ensure the stability of a platoon of more than five vehicles and to avoid "accordion" effects is to use communications so that the leader of the convoy transmits their maneuver intentions to the other vehicles. In this case, all the vehicles in the convoy will be able to react at the same time and not have any accumulation of delay due to the limited range of perception and the inability to be instantly informed of the leader's maneuvers.

Prototyping and evaluating ADAS in a simulation context, a requirement to guarantee quality

Testing and the validation of ADAS on real prototypes and on "open" roads is not always possible due to high material costs, special climatic conditions (desired ones or those to be avoided), or the risks associated with a situation (collision avoidance, mitigation of the impact of a collision, lane departure avoidance, etc.). In addition, it is often very complicated and expensive to be able to obtain reliable and real-time "ground truth". Without this last type of data, the evaluation and validation of ADAS remains qualitative and tedious to implement. In addition, in order to be certified, driving automation services and applications require the generation and use of a very large amount of data and road situations. This is difficult to achieve in real life without driving millions of kilometers. For these reasons, the physico-realistic simulation of road environments (sensors, vehicles, infrastructures, weather conditions and drivers) increasingly appears to be an indispensable and essential tool for enabling the prototyping, testing, evaluation, validation and pre-certifying of ADAS and, by extension, of driving automation. Indeed, these simulation tools will make it possible to generate controlled and repeatable scenarios (such as particular traffic and critical situations) in degraded conditions, with a perfect (ground truth) reference base.

It is obvious that in order to design a complete and operational minimal automated driving application, it is necessary to know how to implement a complete chain of processing steps which involve acknowledging multiple sensors, building local and extended dynamic perception maps using communication tools, understanding and anticipating risky behaviors and situations, fostering optimal decision-making whenever possible, producing both trajectories and path planning, and finally, implementing control laws to generate instructions and orders on the vehicle's actuators. This book intends to address a part of these processing stages. To begin with, the first chapter will address the problem of AI for automated vehicles. Then, the second chapter will discuss how the environment can be perceived through the use of conventional and unconventional cameras. The third chapter will address the problem of trajectory planning. The fourth chapter will present a simulation platform that makes it possible to prototype ADAS and automated driving applications with physico-realistic models of sensors, vehicles and the environment. The fifth chapter will focus on the communication standards for the development of cooperative systems. The penultimate chapter will illustrate the technologies and processing mentioned in the first chapters, with their concrete applications. Finally, the last chapter will address the legal issues underlying the use of Artificial Intelligence for automated driving applications.

References

European Commission (2019). Road safety in the European Union: Trends, statistics and main challenges, April 2018. *26th International Technical Conference on the Enhanced Safety of Vehicles*, Eindhoven, The Netherlands.

Merat, N., Jamson, A., Lai, F., Daly, M., Carsten, O. (2014). Transition to manual: Driver behavior when resuming control from a highly automated vehicle. *Transportation Research Part F: Traffic Psychology and Behavior*, 26, 1–9 [Online]. Available at: 10.1016/j.trf.2014.09.005.

World Health Organization (2019). Road traffic injuries [Online]. Available at: https://www.who.int/ health-236topics/road-safety.

Dominique GRUYER
Gustave Eiffel University
May 2021

Foreword 3

When we get in a car and start driving, we're about to commit to one of the most dangerous activities in our day: driving. And no matter how experienced, careful and respectful a driver is, driving is still the most dangerous portion of our day, because we're sharing the road with other drivers who might not be as experienced, careful or respectful. Plus, when using our car as a means of transportation, driving is not only dangerous but tedious as well since we need to devote our entire attention and energy to this task.

Is there any other task in our average day that is as dangerous, stressful and inefficient? Probably not, and every effort to remove manual driving from our daily routine should be seen as a terrific improvement to our lifestyle.

A vehicle that drives itself does not only bring safety and comfort in people's transportation, but – without exaggeration – can be seen as the next global revolution. The Artificial Intelligence driving the car will be the very first example in human history of an artificial system making safety critical decisions in complete autonomy. Decisions such as overtaking, speeding and negotiating an intersection alongside other vehicles are all maneuvers that may impact not only the passengers' lives, but other road participants like nearby bikers or pedestrians as well, who may not even know that they are interacting with an AI-based system.

Scientists developing autonomous cars need to be aware of their enormous responsibility: a wrong or too quick deployment of a non-completely mature system may reduce people's trust and acceptance of this technology, temporarily blocking the road to AI-based progress in general.

Autonomous driving (AD) technology has extremely powerful potential: it can influence our lives so deeply that it can change our future habits, lifestyle, relationships, space and life expectancy. But unfortunately, turning this highly positive potential into reality is not straightforward. The technical details, implications, downsides and possible misuses of this technology need to be deeply understood not just by scientists, but by lawmakers as well, so that they can influence the way in which this revolutionary technology will be deployed. I hope this book will both guide scientists in their future researches and enlighten governing bodies in making informed decisions on how to best use AD technology for the benefit of our future communities.

Professor Alberto BROGGI

PhD, IEEE Fellow, IAPR Fellow

General Manager, VisLab

May 2021

Preface

The main subject of this book concerns recent developments in Advanced Driver-Assistance Systems (ADAS).

The first chapter will present a global introduction to the concept of Artificial Intelligence, as well as various associated techniques. Then, the problems raised by the application of AI in the industry will be discussed, mainly in the field of autonomous vehicles in relation to ADAS systems.

To begin with, a reminder of the state-of-the-art AI techniques will be discussed in Chapter 1, establishing the link with its applications in the field of autonomous vehicles in the last part of the chapter.

Chapter 2 will be devoted to the perception phase, which resolutely favors visual sensors, albeit conventional or not, due to their low cost and the wealth of data they offer. The first part will present the different types of visual sensors adapted to the problems of autonomous vehicles, while the following sections will be dedicated to the algorithms that exploit the direct outputs of these sensors. The remainder of the chapter will present a selection of techniques showing the way in which low-level approaches integrate the notion of objectives as early as possible. In the field of autonomous systems, objectives require high performance levels: detecting the navigable space, obstacles (vulnerable, two-wheeled vehicles, cyclists, other vehicles) or helping to estimate the ego-vehicle's own movement. Finally, a technique for detecting the navigable space will be presented, as well as its robustness *vis-à-vis* the various disturbances that may affect road scenes, thus introducing a complete visual odometry technique to provide perspective. At the end of the chapter, a projection towards approaches based on deep learning will be evoked, suggesting the possible evolution of these techniques, hand in hand with new AI technologies.

With the development of automated mobility, it becomes necessary to design a certain number of modules, which will make it possible to build a complete automated driving system relying on the data produced by embedded or remote information sources. These modules are perception, decision, and action. The decision-making module will be introduced in Chapter 3. The term "planning" is an old one that reunites the development of theories, models, methods, and approaches applied in robotics, as well as in economics, production and neuroscience. In mobile robotics – close to automated mobility – planning can be broken down into three levels: route planning, trajectory planning, and motion control planning. In this chapter, existing state-of-the-art methods will be presented, as well as the works and applications carried out at Gustave Eiffel University (formerly IFSTTAR), and more particularly, at LIVIC. Among these and other applications, the chapter will present a co-pilot that intends to reach the complete automation of driving, and may possibly interact with the driver.

In Chapter 4, the prototyping, testing and evaluation of ADAS systems will be presented for their effective implementation on the Connected and Automated Vehicle (CAV) of the future.

With the evolution of means of mobility in the last two decades, a large number of ADAS had to be developed. These informative and cooperative applications, which are increasingly active (with the development of automated vehicles), need to be tested, evaluated and validated. However, the stages for assessing the quality, reliability and robustness of these ADAS require the implementation of specific scenarios in a controlled environment, where it is possible to reproduce critical situations (infrastructure degradation, difficult climatic conditions, degradation of the sensor operation, etc.). In addition, in order to assess the performance of these embedded and/or off-board systems, it is imperative to be able to generate "field information" acting as reliable and accurate references. All of these constraints are difficult to take into account and apply in real experiments. In this chapter, an alternative upstream solution to tackle this problem will be discussed. The solution is based on the use of simulation, and more specifically, of the Pro-SiVIC platform. This interoperable, modular, and dynamic platform makes it possible to respond perfectly and efficiently to the constraints imposed during the implementation of an ADAS assessment and validation process. This chapter will discuss the general architecture and various functionalities that need to be implemented in such a simulation platform. In addition, some examples of representative applications which have been prototyped, tested, and evaluated with Pro-SiVIC will also be evoked.

Chapter 5 will be devoted to the communication that makes these systems cooperate with their environment and considerably improve their autonomy.

This chapter will address the question of standardized technologies enabling the exchange of data between vehicles, other road users, road infrastructure, urban infrastructure and traffic and services Cloud Management Platforms. These technologies, standardized to meet European terms by ETSI, CEN and ISO, are known under the name "Cooperative ITS" and bring together several technologies and functionalities for the transfer, organization, securing and processing of data. The best known are undoubtedly short-range localized communication technologies based on a form of Wi-Fi adapted to moving vehicles (ITS-G5). These assist vehicles in communicating directly with one another and with the road infrastructure, without the support of a telecommunications infrastructure (V2X). They are mainly used for applications related to road safety, which ADAS can benefit from. Cooperative ITS also include long-range centralized communication technologies based on the cellular network (LTE, 5G), geo-localized data organization functionalities (*Local Dynamic Map*) and many others which are being standardized. To ensure interoperability, these technologies are grouped together and integrated into a unified communication architecture (*ITS station architecture*) whose motivations, origins, usual cases and vast sets of features will be explained further.

An original case study related to pedestrian detection will be the subject of discussion in Chapter 6. The Moroccan pedestrian represents about 28% of the number of victims of fatal accidents in Morocco. However, these statistics are not shocking considering the unruly behavior of the latter reigning on Moroccan roads. This chapter will present one of the solutions suggested for reducing the rate of pedestrian fatalities in unstructured areas, by proposing a system that mitigates collisions with pedestrians, known as Pedestrian Crash Avoidance Mitigation (PCAM) and based on the detection of the pedestrian's direction, something which has not yet been applied to this type of system. In order to study the Moroccan case closely, a new pedestrian orientation database has been created with data drawn from different Moroccan cities using an integrated camera on a moving vehicle. Pedestrian orientation is detected by a new deep learning algorithms called capsule networks, which have outperformed convolutional neural networks in terms of accuracy.

The regulation of the so-called "autonomous" vehicle is a hot topic both for the general public and for professionals in the sector, such as car manufacturers and innovative companies. The legal impacts induced by the use of autonomous vehicles on roads which are open to traffic are significant. The legislator will soon be confronted with the examination of the consequences and the legal and ethical effects of the generalization of this new technology. In Chapter 7, among other topics, the legal stakes of the use of autonomous vehicles in open environments will be addressed: questions of legal liability and insurance, as well as the General Data Protection Regulation (GDPR), will be discussed.

Abdelaziz BENSRHAIR
INSA Rouen Normandie
May 2021

Artificial Intelligence for Vehicles

1.1. What is AI?

The term "Artificial Intelligence" was coined by John McCarthy [PRO 59] who defined it as "the science and engineering of making intelligent machines." It is worth noting that this definition does not presume the types of methods employed. It is the functional definition of a goal. But this definition mentions the word "intelligent" in the characterization of "artificial intelligence". It is therefore not entirely satisfactory.

A second definition was proposed by Marvin Minsky [MIN 56]: "the science of making machines do things that would require intelligence if done by men." This definition is a little more accurate. Like the first, it is the functional definition of a goal. The term "intelligence" is included in the definition, which again is still not very satisfactory.

Other definitions are commonly used, such as, for example, "a branch of computer science dealing with the simulation of intelligent behavior in computers" and "the capability of a machine to imitate intelligent human behavior"[1], or "the theory and development of computer systems able to perform tasks normally requiring human intelligence, such as visual perception, speech recognition, decision-making, and translation between languages."[2] This last definition offers a little more information about the tasks included within the sphere of intelligence.

Note: "thinking", "spirit", "understanding" and so on are not part of this definition of AI.

Chapter written by Gérard YAHIAOUI.

1 Merriam-Webster American Dictionary, 2020.
2 English Oxford Living Dictionary, 2020.

Nowadays, we consider that the characteristics that make it possible to say whether a technical solution involves AI or not are:

– perception;

– learning;

– reasoning;

– problem solving;

– using natural language.

Here we have a more detailed definition including five verifiable characteristics. If at least one of these features is verified by a system, then – according to this definition – we can say that AI is involved. We observe that learning is only one aspect of what is considered to be AI. This is important because some people systematically associate AI with the ability to learn from data, which leads them to say that the prerequisite for using AI is building large databases. The most commonly accepted definition of AI does not presuppose this, and we may talk about AI systems which do not necessarily learn from data, but integrate knowledge extracted from humans (knowledge-based systems). For years, these systems represented more than 90% of what was called AI, and their weight in current works is increasing, after having almost completely disappeared in favor of learning techniques in recent years. For example, it is now common to again read expressions such as "knowledge representation" in scientific programs.

We can also notice that as technology advances, what was once considered an AI problem is now found in the list of ordinary computer problems, such as for example playing chess: modern computers beat chess champions, not due to any intelligence whatsoever, but because they have the sufficient calculation capacity to play every game sequence combination in parallel virtually and always retain the best sequence. However, for years, this problem was a hot topic for AI researchers.

This remark would tend to show that the notion of intelligence contains an implicit idea of transcendence: as soon as we easily understand how a system works, we no longer consider it to be AI. This leads to three apparently divergent attitudes:

– those who consider that "artificial intelligence does not exist" [LUK 19];

– those considering AI to be a goal which, like the horizon, recedes as we advance;

– those who consider that we will have really created AI when – even perfectly knowing all of its components and parameters – it will not be possible to predict its operation. This is not a far-fetched definition insofar as this notion exists in what is called "deterministic chaos theory" [CRU 86].

In fact, these three attitudes are convergent after all: for those three ways of thinking, the very notion of intelligence must remain transcendent to any explanation reducing AI to a "simple automatism".

1.2. The main methods of AI

This section does not claim to be exhaustive or to present the techniques in detail. It is rather a summary for raising awareness about the existence of the main approaches, in order to understand their specific guidelines and uses.

1.2.1. *Deep Learning*

Deep Learning refers to making a network of "formal neurons" automatically learn a function. Most often, these neurons are made up of a scalar product stage: the input vector is weighted by coefficients called "synaptic weights". Note that when dealing with data vectors, it is customary to use the standard scalar product. But if we process signals (temporal speech signals, images, videos, etc.), we use the scalar product of the signals' Vector Space, which is the intercorrelation function [BUR 19]. Since this function is written as a convolution operation, we then speak of "Convolution Neural Networks" (CNNs) [BUR 19]. This scalar product stage with adjustable coefficients is followed by what is called an "activation function". There are many possible activation functions, the most common being an "S" curve often encoded by a hyperbolic tangent. There are also variants which replace the scalar product with a distance and use non-monotonic activation functions, such as "radial basis functions neural networks" [LOW 88].

In industrial applications, neurons are generally positioned in layers following a "feedforward" pattern. There are two families of layers of neurons: on the one hand, the input layer and the output layer which are in direct contact with the outside world, and on the other hand, the intermediate layer(s), called the hidden or deep layer (hence the term Deep Learning).

These neural networks have a particular interest: first of all, their operation is stable because they have no direct or indirect feedback from a neuron's output in relation to its inputs, and second, there is a theorem showing that – under certain conditions which the activation function must respect – there are universal approximators [HOR 89].

The learning of such layer networks involves iteratively adjusting the variable parameters of the scalar products, so as to minimize the error made by the neural network on what is called a "resolved" database (or learning base). We speak of supervised learning, because during the learning process, for each input of the neural network, we know its desired output.

This iterative search for the possible minimum of the error made by the network uses classical concepts of operational research, such as random search, and sophisticated features like simulated annealing [HEN 06], or the notion of the error slope with gradient descent leading to what was called the "backpropagation" learning rule [BUT 05].

There are many other types of neural networks, but multi-layer feedforward neural networks are the most commonly used.

1.2.2. *Machine Learning*

In general, a distinction is made between Machine Learning and Deep Learning. Deep Learning is the expression of Machine Learning in the particular case of multi-layered feedforward neural networks.

The framework of Machine Learning is broader: whatever the constituent parts of a system, we can represent them as being a calculation system or by a logical reasoning element whose parameters can be automatically adjusted. The iterative search for parameters to minimize an error on a database is called Machine Learning, and employs the same concepts of operational research.

Machine Learning systems quite often include:

– Bayesian networks: they manipulate "Bayes" conditional probability calculation cells [PEA 88];

– transfer learning systems: this involves reapplying knowledge and skills initially used to solve one task, in order to solve another one. The difficulty lies in measuring task proximity and in knowing which characteristics of the task are related to what kind of knowledge [BOZ 76];

– fuzzy decision tables [GUP 82]: these implement the classic notion of the decision table by replacing the classic "yes/no" logic by fuzzy logic which admits all the degrees between "yes" and "no" via an operator membership function to a fuzzy set, which is a configurable continuous function. Machine Learning involves iteratively finding the adjustment parameters of fuzzy membership operators in order to minimize decision errors on a database. Note that fuzzy sets can be used to represent both uncertain knowledge and imprecise data and often use "possibility theory" [DUB 88] to compute a decision;

– decision trees [KAM 17]: the decision space is iteratively split into groups of cases, and each one is split into smaller groups, etc. The iteration of this segmentation process builds a decision tree. Each tree node has adjustable parameters which make it possible to classify a case into one group of cases or into another. The iterative search for node parameters minimizing decision errors on a database also falls under the scope of Machine Learning.

There are many other Supervised Machine Learning techniques, but the above-mentioned summary is enough to understand the rest of the book.

1.2.3. *Clustering*

One of the questions for understanding a large volume of data is the following: is it possible to divide the database into so-called homogeneous groups? Groups are said to be homogeneous if the intra-group variance is smaller than the inter-group variance. This question, also called "vector quantization" [PAG 15] has led to the development of many methods, among which we can mention:

– Moving average or K-means clustering [LLO 57]: this algorithm is simplistic. It involves initializing "seeds" (random vectors in the space of the data vectors), and assigning to each seed the vectors which are closer to one seed than to the other seeds. Thus, we obtain a group of vectors per seed. New seeds are calculated as being the barycenters of each group and the operation is repeated N times. After a while, the seeds stop moving, and the algorithm is said to have converged. Note that this algorithm was developed outside the AI sector by researchers working on Data Analysis and Signal Theory, and so moving average is not customarily considered to be part of the AI techniques. But this method is still the basis used by researchers in automatic clustering neural networks, which use this simple idea but implement a stochastic version into an iterative learning rule to improve performance.

– At first glance, the Self-Organization Maps (SOM) proposed by Teuvo Kohonen [KOH 82] can be seen as a Monte-Carlo [MET 49] K-means method using a neural network. Each neuron has its vector of coefficients (called synaptic weights) that operates a ponderation of input. By construction, this vector has the same number of components as the space to be divided into groups. When we have an example to be classified, we locate the neuron whose vector of coefficients is closest to this input vector. This neuron is called the "elected neuron". The input coefficients of the selected neuron are modified so that they are even closer to the input example. We thus present the set of examples to be classified N times. The idea is that when we bring the coefficients of the selected neuron closer to an example input, we also bring it closer to all the data that looks like this input vector.

A moving average is thus implemented without ever calculating the barycenter, something which offers mathematical advantages. But in order for this to become a SOM, we geographically position the neurons on a (usually 2D) map and apply an interaction between neurons – called lateral inhibition – inspired by the structure of the biological neural networks of our perception. The key point is to consider that the geographical neighbors of the elected neuron also have to modify their synaptic coefficients, but following a function called a "Mexican hat": the near geographical neurons also bring their synaptic coefficients closer to the input vector which has contributed to choose this selected neuron, but they do so with decreasing force due to their distance from the selected neuron. On the contrary, geographically distant neurons have their synaptic coefficients far from the input example. At the end of this process, each zone of the map is sensitive to a type of input, and the classes are geographically positioned depending on the topology: two neighboring classes are represented by groups of close neurons.

– Neural gas [MAR 91]: these are an extension of SOMs. Instead of measuring the distance between neurons in a geographic space, we calculate this distance within the space of synaptic coefficients. In other words, we consider that the closest neuron to the elected neuron is the neuron which would have been elected had the elected neuron not been there. This establishes a distance which sorts neurons differently for each input example, as if neurons were constantly moving in relation to one another, which explains the term "neural gas".

At the end of the clustering operation, we obtain homogeneous groups of data without ever having to say which example fell within each group. The examples are grouped according to a similarity principle ("birds of a feather flock together"). It is for this reason that we speak of unsupervised learning.

1.2.4. *Reinforcement learning*

Reinforcement learning can be compared to supervised learning, indirectly: we do not know the desired output, but we know whether the computed output is admissible or not. The idea is to discourage unacceptable solutions and to reinforce the presence of acceptable solutions. This is called a cumulative reward mechanism.

Note that we also find this kind of idea of "admissible" solutions in other fields, such as in automation, for example, using the viability theory [AUB 11] for servo-control problems. This involves not seeking an optimal output depending on an optimization criterion, but a solution enabling the system to be "viable", that is to say, acceptable from the point of view of piloting (a robot, for example).

Most of the classical reinforcement learning methods suffer from a combinatorial problem and are only suitable for problems with few dimensions (5 maximum for example). For instance, let us quote the Markov Decision Process (MDP) [OTT 12] serving as Q-Learning or dynamic programming support: we consider that the system to be solved takes N number of states modeled by a Markovian process.

In the case where we do not know the process, we can use Q-learning [SUT 98]. This involves learning a decision (a robotics action) for each state of the system. During the learning process, we make a Q (a, b) function evolve. This represents the (observed) reward given to action "b" when the system is in the "a" state. When operating, the system applies the action which maximizes the reward in the S state in which the system is set.

If we know the MDP, we can use dynamic programming [BEL 57], an optimization method dating from the 1950s, stipulating that the optimal response to a problem is a combination of the optimal responses to subproblems. "Divide and conquer", as they say.

For problems with a large number of dimensions, it is possible to use:

– Deep Learning, as presented above, and in that case, we speak of Deep Reinforcement Learning (DRL) [MAT 15].

– Genetic Algorithms [HOL 84]: a set of solutions is considered as a population of individuals represented by a data vector which is called a genotype. Each genotype (potential solution) gives rise to a result called a phenotype. A cumulative reward function makes it possible to kill or not to kill individuals. Individuals can reproduce in pairs by crossing each other (crossing over), which amounts to copying and pasting the ends of genotypes to make new individuals, and individuals can mutate (random modification of an element of the genotype vector).

The mechanism is Darwinian: the best solutions survive and are the only ones to contribute to reproduction in view of creating new solutions. Unlike Deep learning, we do not iteratively modify a solution to minimize an error, but iteratively modify a population of solutions in order to maximize the acceptability of these solutions. We therefore try to develop many solutions at the same time by crossing them, so as to find the best ones.

It is for this reason that we speak of "implicit parallelism". The use of this approach requires having coded the genotypes using a coding for which at least one local topology exists so that the interpolations between solutions are meaningful.

1.2.5. *Case-based reasoning*

Case-based reasoning [AAM 94] draws its inspiration from the resolution of similar problems. For this, it is necessary to have a description of the problems which accept a topology in order to be able to calculate the distance between cases: if two representations are similar, the problems should be close, and vice versa.

The search for similar problems is generally performed heuristically, or quite simply by searching for the N closest neighbors.

The underlying principle is to record the reasoning which made it possible to successfully solve a case, and then, either to directly apply the reasoning behind the closest solved case, or to apply a modification (a derivative) of the reasoning behind the closest cases.

The system continues to learn by storing more and more cases with the reasoning that made it possible to solve them.

1.2.6. *Logical reasoning*

Knowledge-based systems generally seek to retrieve knowledge from human experts capable of solving complex problems through logical deduction. Before knowledge-based systems are ready for use, it is necessary to elicit knowledge from the experts, by means of interviews and the resolution of study cases. When this logical knowledge is drawn, it needs to be modeled (as knowledge representation [SOW 00], generally in the form of mathematical logic).

To solve a problem, logical reasoning is applied by a special program called an inference engine which can operate with several levels of complexity: it can make assumptions, validate or invalidate them, manipulate logic using all of "de Morgan's rules", etc. The logic used can be binary or fuzzy [ZAD 75]. In the latter case, we can use more recent mathematical methods in order to combine logical operators with the theory of possibilities [DUB 88], for example.

1.2.7. *Multi-agent systems*

Multi-agent systems [WEI 99] are inspired by groups of animals functioning in swarms: bees, ants, etc.

Each agent (a bee) has a very simple behavior, like an automaton, but it is the interactions between automatons which endows the group with what is called an emergent property [OCO 12] making it possible to solve a complex problem.

An agent has autonomy of action and a reaction model towards its environment, following a "Belief-Desires-Intentions" (BDI) type [RAO 95]. The key to multi-agent functioning is communication between agents as well as their interaction (influence of the action of an agent on the actions of nearby agents, for instance). Learning in multi-agent systems resorts to known methods like the Markov Decision Process (MDP), among others.

1.2.8. *PAC learning*

PAC (Probably Approximately Correct) learning [VAL 84] introduces complexity theory into the learning process. The idea is that we have pre-established classification functions, associating a class with each input vector.

One of these pre-established functions is called a "concept", and is the (complex) function responsible for properly classifying new entries.

In practice, we can initialize a collection of functions. For that, we have to find the function which is the closest to the concept in polynomial time, that is to say, the function which has a higher probability than P of providing the concept, with an error smaller than E.

1.3. Modern AI challenges for the industry

1.3.1. *Explainability: XAI (eXplainable Artificial Intelligence)*

AI techniques may work very well without their output being easy to understand, even when performing an *a posteriori* audit. However, in certain areas involving discrimination between individuals (decision to award credits, for example), or other areas with strong safety stakes (driving an autonomous vehicle on an open road, for example), transparency requirements are demanding. In the first case, transparency or "explainability" makes it possible to justify a decision, whereas in the second case, it helps produce a formal security proof, for example by using the SIL (Security Integration Level) formalism [THE 19].

Those works which make the functioning of AI systems explainable are grouped under the term XAI [GUN 19] (eXplainable Artificial Intelligence).

One should note that – for this subject – knowledge-based systems have an advantage over purely numerical computing methods, such as Deep Learning, for instance. The dilemma when retrieving an XAI is that one generally must accommodate a decrease in performance in order to gain better explainability.

1.3.2. *The design of so-called "hybrid" AI systems*

Apart from a few experts in a given technique who try to solve all the problems with that technique at all costs, the engineers who have to design AI systems have come to the conclusion that it is necessary to use a collection of several techniques, in other words, a "hybrid" AI solution.

This therefore leads to solutions which can jointly integrate:

– Deep Learning;

– knowledge-based systems;

– stats;

– physics;

– applied mathematics;

– classic algorithms (image processing, signal processing, etc.).

The design of such a solution requires having broad general knowledge, and to avoid drowning in the almost infinite combination of choices.

For this, a methodology has been created to design solutions mixing classical algorithms with neural networks. It is called "AGENDA" (*Approche Générale des Etudes Neuronales pour le Développement d'Applications*, General Approach to Neural Studies for the Development of Applications) [AMA 00]. This methodology considers the system to be designed as an information processing machine, specified by its variants and invariants. In practice, it makes it possible to facilitate team creativity (it is necessary to pool the skills of several experts on various techniques), so as to obtain the traceability of technical choices, improve testing procedures, and – as far as learning is concerned – qualitatively and quantitatively polish learning and validation databases.

One of the effects observed from the application of this method is that one can obtain more efficient and more robust learning with a much smaller number of examples.

1.4. What is an "intelligent" vehicle?

In the past, humankind has already had access to a mobility vehicle with its own intelligence: the horse.

But it is clear that the piloting of this vehicle can sometimes be complex, namely when the assessment of the situation made by the horse and by the rider, differs.

When we speak of an intelligent vehicle, we exclude the case of vehicles having their own free will, except in specific cases where we consider that the driver is not able to properly assess the situation.

This is called driving assistance, or delegation of driving, which can either be partial or complete.

1.4.1. *ADAS*

The purpose of ADAS (Advanced Driver Assistance Systems) [HER 16] is to inform the driver of situations he/she might not have perceived, but above all to act on the vehicle automatically in order to help the driver by acting in his/her place. We can divide the field of ADAS into two parts:

– the action while driving normally: there is no emergency, the vehicle is traveling normally in a suitable environment, and the vehicle can, for example, automatically regulate its speed (Intelligent Speed Adaptation-ISA [BLU 12]) via a device called Automated Cruise Control (ACC) [WIN 12]. We can consider that this constitutes a delegation of the longitudinal part of driving. Likewise, when the driving lane ahead of the vehicle can be clearly detected, the vehicle can automatically control the steering wheel in order to remain at the center of this lane, using Lane Departure Control and steering assist [GAY 12]. We can consider this function as a delegation of the lateral part of driving.

– Emergency action: when the vehicle detects an emergency situation, it is usually too late to alert driver (this can even be counterproductive because it generates disruptive stress at a dangerous moment). There are several phases: during the emergency, during crash, after crash.

1.4.1.1. *The different driving phases*

1.4.1.1.1. Normal driving

The system is expected to "perform well", seeking to minimize the occurrence of an emergency situation. As seen above, this is the domain of ACC (Automated Cruise Control) and "Lane Departure Control".

1.4.1.1.2. Emergency situation

If possible, we want the system to avoid the accident by reacting to the emergency situation with a quick and appropriate reflex. In the worst scenario, if the accident is inevitable, we want the system to make it the least serious possible. In this category, we can find:

– the vehicle's dynamics control systems: brake assist and traction control;

– automatic braking systems: AEB (Automatic Emergency Braking) [AEB 17];

– potentially, automatic avoidance systems. Since its implementation is likely to generate an accident with the vehicles in the adjacent lane – surprised by the emergency avoidance – this action is more delicate than emergency braking.

1.4.1.1.3. During the accident

If the accident cannot be avoided, the vehicle offers solutions to minimize its severity: deformation zone, safety capsule, airbag, seat belt.

1.4.1.1.4. After the accident

Depending on the estimated severity of the shock, an automatic telecommunications system can decide to call for rescue.

1.4.1.2. *Highly distinct notions related to the idea of danger on the road*

When it comes to potential accidents in the future, all engineers instantly think about the notion of "probability of collision". Of course this notion is important, but, as we will see below, other notions are important too.

1.4.1.2.1. Potential danger ("hazard")

For a long time, potential danger has been signaled by road signs like, for example, "ice hazard" or "rockfalls". The disadvantage of these signs is that it is not always clear whether they make sense in all driving conditions. Nowadays, alerts can be integrated on variable message panels, or directly in the cockpit of the vehicle in the form of sound or voice alerts. Some such alerts are: accident in 3 km, traffic jam in 4 km, people walking in the lane, etc.

This information regarding potential danger is either integrated into the digital map, or sent to the vehicle from other vehicles, or from the road infrastructure using a telecommunications system, generally called the V2X system (vehicle to vehicle (or infrastructure) telecommunications, coining the phrase "Connected Vehicle").

This information is interesting, but in practice, the driver often does not know what to do or when to do it.

The hazard [DUD 17] will only become an effective danger if the driver's driving behavior is unsuitable.

1.4.1.2.2. Driving risk

When one monitors both the driving behavior (speed, acceleration) and the driving context (complexity of the infrastructure ahead, hazard warning, traffic conditions, etc.), any mismatch between these two elements can be detected. Such a mismatch is considered as a lack of caution in "driving risk" [GRÉ 16]. This lack of caution/driving risk can be estimated in real time using knowledge-based systems. Driving risk is estimated continually during normal driving. Its level determines the frequency (low or high) with which a vehicle can find itself in an emergency situation (emergency scenarios being the trigger for an accident situation). As part of the above-mentioned ISA problem, an ACC can automatically regulate the speed of the vehicle in order to prevent speeding, of course, but also to prevent emergency situations that may lead to an accident.

1.4.1.2.3. Criticality

The criticality of an emergency situation is the effective level of danger [PAU 18] during an emergency scenario. This level can vary very quickly over time. Criticality can be estimated via several indicators such as:

– probability of collision (or probability of trajectories interception);

– time to collision (TTC);

– probable shock energy.

Note that common natural language makes it possible for the word "risk" to be used quite naturally in these three cases (potential danger, lack of caution/driving risk, criticality). As a result, this makes some communications incomprehensible. Here, we prefer to distinguish the concepts by using the appropriate term each time.

1.4.1.2.4. Global diagram of ADAS

A modern vehicle equipped with driving assistance can be seen as an automatic device including two features: a branch dedicated to normal driving, with ACC, for example (or the Lane Departure System), and a second, faster branch, dedicated to ADAS for emergency situations, such as AEB.

The normal driving branch focuses on the extended ISA (Intelligent Speed Adaptation) aspect with automatic control of speed, which seeks to reduce the vehicle's exposure to risk (keeping the driving behavior cautious). It is in this branch that we will come across the problem of respecting speed limits and that we will first have to estimate the driving risk, so as to keep it at an acceptable value.

The emergency management branch continuously assesses criticality. When criticality exceeds a severity threshold, a switch device enables switching to emergency mode, whereas if criticality is zero or very low, the switch continues in normal driving mode.

The robot (ACC for example) executes the command in all cases.

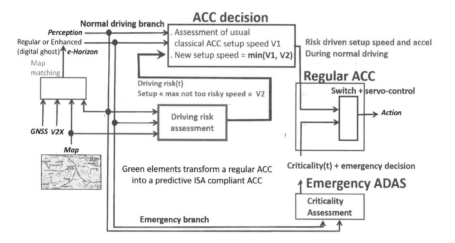

Figure 1.1. *Global diagram of ADAS. For a color version of this figure, see www.iste.co.uk/bensrhair/adas.zip*

Note that this architecture makes it possible to implement an extension of the ISA functionality (automatic speed control) as well as to guarantee the control of risk-taking, without modifying the architecture of the classic ACC. It only requires integrating the function for calculating the driving risk (in green in Figure 1.1), which also includes the speed limit (read on the map or captured by perception).

1.4.2. *The autonomous vehicle*

1.4.2.1. *Different missions*

Autonomous vehicles have been highly publicized, but in the end one can realize that this term covers different types of missions which have often been quite shallowly explained:

– The autonomous bus: a normal bus always runs at the same times on the same route (the bus line's route). This route and its characteristics include invariants which can be "learned by heart". There still may be environmental variability since sunshine and other weather conditions are likely to modify the outputs of the perception sensors, for the same place. In addition, temporary, moving objects (works on the road, for example) can modify the perceived local geometry. But these variations are few in number and easy to supervise or even "learn", for instance in the case of sunshine and weather conditions. The traveling objects of this type which are currently being deployed are minibuses, called "autonomous shuttles".

– The autonomous taxi: the autonomous taxi as currently envisaged makes it possible to go from one known locality to another known locality, on demand. This represents additional complexity since the potential routes are all those which go from one locality to another. If we have N localities, we have N x (N-1) routes. We can refer back to the case of the bus and "learn by heart" all the possible routes, even if N is relatively large (for example N = 100 produces 100 x 99 = 9,900 possible routes). We can also restrict the taxi's radius of action (example: Paris or Los Angeles), and memorize every street in both directions (in the case when both directions are allowed), in which case the taxi can go from any address to any address within this restricted perimeter, instead of connecting only a limited number of localities. In a city like Paris which has about 5,000 streets, this yields an order of magnitude of 10,000 streets to be learned by heart, in terms of the sensors' expected outputs: for example, we can have images at "every" place on every street, with no moving elements (deserted streets), and at each moment compare the image acquired in real time with the corresponding reference image: all the differences between these two images are object leads (except the shadows cast in sunny weather and other such disturbances).

– The autonomous personal vehicle: one might think that the autonomous personal vehicle looks like a taxi. But a personal vehicle has no range limitation: a driver from Canada can decide to go to Brazil by car, to a specific address. In this case, we can clearly see that it would be difficult to imagine having precise references for each place of the journey. We should all the same note that digital map suppliers wish to make this type of reference with a 3D map with an accuracy close to 5 cm. But building such a reference at the global level is a very long process. This means that, for the time being, this particular autonomous vehicle must be able to drive in an unfamiliar environment (just like a human being). The level of complexity then increases considerably, which explains the delay in achievements compared to the announcements made by manufacturers a few years ago.

Let us observe that a partial solution could be assimilated to the case of the taxi: in Europe, for those of working age, 56% of the distance traveled per year is between home and work. One could imagine a vehicle that learns the references of the home–work journey by heart, then the children's home–school trajectory, etc. in order to complete 90% of the journeys independently, like a shuttle. But most car manufacturers have chosen to break down journeys, not into mobility functions, but into infrastructure types: highways, roads, cities. The vehicle must therefore move properly in an unfamiliar environment, as well as cope with the ample variability of infrastructures. It is therefore easy to understand why autonomy functions start with highways, the simplest and most "standardized" type of infrastructure.

1.4.2.2. *Different levels of autonomy*

Autonomy levels are divided into five levels:

– Level 1

At level 1, the driver is still in charge of most of the car's driving functions, but with a little assistance. For example, a Level 1 type of vehicle may provide brake assist if you get too close to another vehicle, or may have an adaptive cruise control function to control inter-distance and highway speed.

– Level 2

Level 2 vehicles may have automatic systems which take charge of all the aspects of driving: steering, acceleration and braking. But the driver must be able to take control of the driving in the event of a failure of part of the system. Level 2 could therefore be seen as "with no intervention", but the driver must in fact always keep his hands on the wheel.

– Level 3: conditional autonomy

Conditional automation allows drivers to sit back and let the car do all of the driving, with no eye involvement. The driver can focus his attention on other activities. But this functionality is only available in certain cases (for example in traffic jams or on the highway). When we get out of these conditions, the driver must regain control, he must therefore stay awake.

– Level 4

Level 4 cars are called "mind-off" vehicles because the driver is not required at all, to the point that the driver can even go to sleep. But there are some restrictions, because the autonomous driving mode can only be activated in the certain areas, or in case of traffic jams and other well-defined cases.

– Level 5

Level 5 does not require any human interaction. The vehicle can steer, accelerate, brake and monitor road conditions, enabling the driver to sit down without having to pay any attention to the car's functions.

As explained above, the complexity of the achievement is not only related to the autonomy level, but also to whether the environment is entirely known or not, or on the contrary, is completely unknown. In this sense, an autonomous shuttle at level 5 is only able to complete one trajectory. This ultimately looks easier to achieve than making a level 3 vehicle go from one place to another at any moment.

1.4.2.3. *Liability in the event of an accident*

From the moment that it is no longer the driver who drives, but the vehicle itself, the question of liability for potential accidents arises. In the event of an accident, it might be interesting to have a "black box" having recorded signals clearly showing whether the vehicle had been driven carefully or not. This corresponds to the concept of driving risk (or lack of caution), as defined above. The use of such a system for monitoring caution or the lack of caution at the time of the accident could be one lead for dealing with the issues of legal liability.

1.4.2.4. *General diagram of the autonomous vehicle*

We can draw a diagram very similar to that of ADAS: the "emergency" part is very similar: the purpose is to detect emergency situations. The most important part in terms of complexity is the normal driving branch, since the vehicle must follow the road's rules under all conditions, and respect the driving cautiously rules, adapted to the situation.

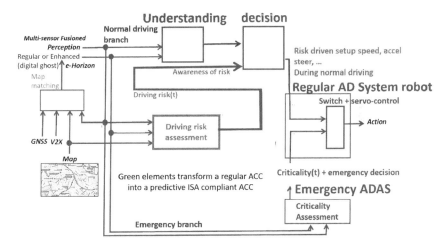

Figure 1.2. *Global diagram of ADAS. For a color version of this figure, see www.iste.co.uk/bensrhair/adas.zip*

Perception is almost always muti-sensor (several cameras and/or lidar, radar, etc.). Compared to predictive ACC, it is no longer just a question of defining a safe driving speed, but of managing the vehicle's entire driving behavior (speed, acceleration, direction, etc.). The "understanding" part is generally based on the concept of "situation holding": at a given moment, we have a representation of the world with the infrastructure, all detected objects (their nature, relative speed, direction, etc.), the detection's reliability, the context (day, night, fog, etc.), and so on. We predict the evolution of this representation at the following instant. If the predicted situation is sufficiently close to the acquired situation, then the situation is considered to be "well understood". For objects which have been detected and disappear without being able to see them leaving the sensors' field, we generally continue to make them evolve as ghosts in the representation of the world, with a potential oblivion over time. For example, this would be the case of two wheels remaining at a possible blind spot. If the prediction turns out to be seriously false, this means that the representation of the world is erroneous (we then speak of "hypothesis breaking"), and we are confronted with an incident which is generally reported to the branch, and possibly triggers specific monitoring.

1.4.2. The construction of the intelligent vehicle's basic building blocks employing AI methods

This chapter presents application examples of AI techniques to the basic building blocks making it possible to build the diagrams presented above. We do not claim to be exhaustive, but rather illustrative. In point of fact, in most engineering cases, there are no unique solutions.

1.4.2.1. Normal driving management

1.4.2.1.1. The purpose of perception bricks in normal driving

In normal driving conditions, the sensors (camera, lidar, radar, etc.) are not intended to detect emergency situations in order to estimate their criticality, but rather to estimate how the driving complies with the rules of caution by extracting the risk factors from the scene. A good illustrative example is the measurement of the distance from the vehicle ahead: if two vehicles both travel at exactly the same speed of 130 km/h, and they follow each other keeping to a two meter distance (inter-distance = 2 m), then one understands that the follower (vehicle) is not being careful, in other words, is running a risk. But criticality is low, almost zero: the probability of collision is zero and the time to collision is infinite since the two vehicles are both traveling at the same speed.

Conventionally, perception sensors are cameras, radars, and lidars. These sensors provide signal streaming and real-time images. We will not focus on sensors here because the focus of this chapter is the use of AI techniques (which intervene after sensors do).

Note that perception in the normal driving mode also has the objective of feeding the "situation holding" brick, in order to check that the understanding of the scene is correct.

The AI technique most used for this perception brick is Deep Learning.

The main applications are:

– detection of other vehicles;

– detection of vulnerable people (pedestrians, bicycles, etc.);

– detection of weather conditions (rain, fog, snow, etc.), and in particular their impact on visibility [YAH 03].

In the images obtained from cameras, in particular, we find real applications of CNNs, and other applications of hybrid systems combining algorithms for retrieving image features, using multilayered neural networks.

1.4.2.1.2. Understanding brick

The understanding and situation holding brick as described above can use different techniques: of course, the whole "prediction" part can be carried out by logical reasoning propagating and validating or invalidating hypotheses to achieve consistency. This part is still embryonic in many vehicles which, for the moment, are limited to classifying the data of integrated sensors in one class. The class number is named by a typical life situation on the road. But it is important to check the spatial and temporal consistency of this life situation so as to really speak of "understanding".

1.4.2.1.3. Decision brick

Decision-making can involve:

– logical reasoning: the advantage is the transparency of the system's operation. The downside is that it is difficult to find a method for representing life situations on the road which is easy to handle and sufficiently comprehensive;

– a purely numerical calculation method such as neural networks is simpler to implement, but is not conducive to the transparency of the decision-making process.

To train a (symbolic or digital) decision system, we are in the typical case of reinforcement learning: if the decision leads to a road safety situation which is acceptable, then one reinforces this decision, otherwise one penalizes it. After learning, the decision system always remains in construction for acceptable decision cases. At least, this is the goal that must be validated.

We must estimate acceptability: as we have explained, there is a brick called "driving risk assessment" which has been precisely designed for this. This function can be used during reinforcement learning, in real time as in control, or during validation tests.

1.4.2.1.4. Driving risk estimation

Driving risk is the degree to which driving behavior is unsuitable for the road context. We interpret this degree of mismatch as a lack of caution which coincides with the notion of the risk triangle proposed by Frank E. Bird and H.W. Heinrich [HEI 31]. Today there is a construction of such a brick [BRU 19], which gives a new impetus to the autonomous vehicle. It is a hybrid AI, mainly based on a knowledge-based progressive logic system (running deep knowledge in road safety), which – among other things – employs the theory of possibilities, and apart from that, Deep Learning for the recognition of life situations on the road.

1.4.2.1.5. Potential danger alerts

The potential danger warning (or "hazard warning") is a direct application of digital geographic maps and telecommunications (V2X). This involves informing the vehicle of potential dangers (fog, accident, etc.). We should note that these alerts can be used as an entry for the estimation of the driving risk, because they characterize the driving context. The approach by the vehicle towards a potential hazard will or will not be transformed into a driving risk, depending on the driving behavior.

1.4.2.2. *Emergency management*

1.4.2.2.1. The purpose of perception bricks in normal driving

As presented above, emergency situations are characterized by their level of criticality (probability of interception of trajectories, time to collision, etc.). Emergency is generally defined as "less than 1.4 seconds to collision if nothing is done". The goal is to apply a safe and automatic reflex very quickly, such as the AEB (automatic emergency braking).

The elements detected are the same as during normal driving, but the information retrieval that is made from it is different, since we focus on the calculation of criticality instead of the driving risk calculation (for example, the "time to collision" instead of the "inter-distance").

1.4.2.2.2. A short chain towards decision-making: the concept of reflex

In the case of reflexes, decision is directly connected to perception, as happens with human beings (example: rapid closing of eyes in the case of rapid movement in the vicinity). There is usually no elaborate step of understanding because the emergency requires immediate action. Most systems are based on a ranking of perception data that tells what to do. This is the preferred field of neural networks and Deep Learning.

1.5. References

[AAM 94] AAMODT A., PLAZA E., "Case-based reasoning: foundational issues, methodological variations, and system approaches", *Artificial Intelligence Communications*, vol. 7, no. 1, pp. 39–52, 1994.

[AMA 00] AMAT J.L., YAHIAOUI G., *Techniques avancées pour le traitement de l'information : réseaux de neurones, logique floue, et algorithmes génétiques*, Cepaduès, 2000.

[AUB 11] AUBIN J.P., BAYEN A.M., SAINT-PIERRE P., *Viability Theory: New Directions*, Springer Science & Business Media, Berlin, Heidelberg, 2011.

[BEL 57] BELLMAN R., *Dynamic Programming*, Princeton University Press, Princeton, NJ, 1957.

[BLU 12] BLUM J.J., ESKANDARIAN A., ARHIN S.A., "Intelligent Speed Adaptation (ISA)", in ESKANDARIAN A. (ed.), *Handbook of Intelligent Vehicles*, Springer, London, 2012.

[BOZ 76] BOZINOVSKI S., FULGOSI A., "The influence of pattern similarity and transfer learning upon training of a base perceptron", *Proceedings of Symposium Informatica*, 3-121-5, Bled, 1976.

[BRU 19] BRUNET J., DA SILVA DIAS P., YAHIAOUI G., *Real Time Driving Risk Assessment for Onboard Accident Prevention: Application to Vocal Driving Risk Assistant, ADAS, and Autonomous Driving*, SIA CESA 2018 – Electric Components and Systems for Automotive Applications, Components and Systems for Automotive Applications, 2019.

[BUR 19] BURRUS C.S., *Vector Space and Matrix in Signal and System Theory*, Rice University Publishing, Houston, TX, 2019.

[BUT 05] BUTZ M.V., GOLDBERG D.E., LANZI P.L., "Gradient descent methods in learning classifier systems: Improving XCS performance in multistep problems", *IEEE Transactions on Evolutionary Computation*, vol. 9, no. 5, pp. 452–473, 2005.

[CHA 16] CHANDAR S., KHAPRA M.M., LAROCHELLE H. *et al.*, *Correlational Neural Networks*, *Neural Computation*, MIT Press, Cambridge, MA, 2016.

[CRU 86] CRUTCHFIELD T., MORRISON J.D., FARMER P. *et al.*, "Chaos", *Scientific American*, vol. 255, no. 6, pp. 38–49, 1986.

[DUB 88] DUBOIS D., PRADE H., *Possibility Theory: An Approach to Computerized Processing of Uncertainty*, Kluwer Academic/Plenum Publishers, New York/London, 1988.

[DUD 17] DUDZIAK M., LEWANDOWSKI A., ŚLEDZIŃSKI M., "Uncommon road safety hazards", *Procedia Engineering*, vol. 177, pp. 375–380, 2017.

[GAY 12] GAYKO J.E., "Lane departure and lane keeping", in ESKANDARIAN A. (ed.), *Handbook of Intelligent Vehicles*, Springer, London, 2012.

[GRÉ 16] GRÉGOIRE J., DA SILVA DIAS P., YAHIAOUI G., "Functional safety: On-board computing of accident risk", *Advanced Microsystems for Automotive Applications*, pp.175–180, 2016.

[GUN 19] GUNNING D., STEFIK M., CHOI J. *et al.*, "XAI–Explainable artificial intelligence", *Science Robotics*, vol. 4, no. 37, 2019.

[GUP 82] GUPTA M., SANCHEZ E., *Approximate Reasoning in Decision Analysis*, North-Holland Publishing Company, Amsterdam, 1982.

[HEI 31] HEINRICH, H.W., *Industrial Accident Prevention: A Scientific Approach*, McGraw-Hill, New York, 1931.

[HEN 06] HENDERSON D., JACOBSON S.H., JOHNSON A.W., "The theory and practice of simulated annealing", in GLOVER F., KOCHENBERGER G.A. (eds), *Handbook of Metaheuristics*, vol. 57, Springer, Cham, 2006.

[HER 16] HERMANN WINNER H., HAKULI S., LOTZ F. *et al.* (eds), *Handbook of Driver Assistance Systems, Basic Information, Components and Systems for Active Safety and Comfort*, Springer, Cham, 2016.

[HOL 84] HOLLAND J.H., "Genetic algorithms and adaptation", in SELFRIDGE O.G., RISSLAND E.L., ARBIB M.A. (eds), *Adaptive Control of Ill-Defined Systems. NATO Conference Series (II Systems Science)*, vol. 16, Springer, New York, 1984.

[HOR 89] HORNIK K., STINCHCOMBE M., WHITE H., "Multilayer feedforward networks are universal approximators", *Neural Networks*, vol. 2, no. 5, pp. 359–366, 1989.

[KAM 17] KAMIŃSKI B., JAKUBCZYK M., SZUFEL P., "A framework for sensitivity analysis of decision trees", *Central European Journal of Operations Research*, vol. 26, no. 1, pp. 135–159, 2017.

[KOH 82] KOHONEN T., "Self-organized formation of topologically correct feature maps", *Biological Cybernetics*, vol. 43, no. 1, pp. 59–69, 1982.

[LLO 57] LLOYD, S.P., Least square quantization in PCM, Paper, Bell Telephone Laboratories, 1957.

[LOW 88] LOWE D., BROOMHEAD D.S., "Multivariable functional interpolation and adaptive networks", *Complex Systems*, vol. 2, pp. 321–355, 1988.

[LUC 19] LUC J., *L'intelligence Artificielle n'existe pas*, Éditions First, Paris, 2019.

[MAR 91] MARTINETZ T., SCHULTEN K., "A 'neural gas' network learns topologies", *Artificial Neural Networks*, Elsevier, Amsterdam, 1991.

[MAT 15] MATIISEN T., Demystifying Deep Reinforcement Learning, Computational Neuroscience Lab, available at: neuro.cs.ut.ee, 2015.

[MET 49] METROPOLIS N., ULAM S., "The Monte Carlo method", *Journal of the American Statistical Association*, vol. 44, no. 247, pp. 335–341, 1949.

[MIT 56] MASSACHUSSETS INSTITUTE OF TECHNOLOGY, Heuristic Aspects of the Artificial Intelligence Problem, MIT Lincoln Laboratory Report 34-55, ASTIA Doc. No. AS 236885, 1956.

[OCO 12] O'CONNOR T., WONG H.Y., "Emergent properties", *The Stanford Encyclopedia of Philosophy*, Stanford University, Stanford, CA, 2012.

[OTT 12] VAN OTTERLO M., WIERING M., "Reinforcement learning and markov decision processes", *Reinforcement Learning, Springer*, Cham, 2012.

[PAG 15] PAGÈS G., "Introduction to vector quantization and its applications for numerics", *Proceedings and Surveys, EDP Sciences*, vol. 48, no. 1, pp. 29–79, 2015.

[PAU 18] PAULSEN C., BOYENS J., BARTOL N., Criticality analysis process model: Prioritizing systems and components, NIST report NISTIR 8179, 2018.

[PEA 88] PEARL J., *Probabilistic Reasoning in Intelligent Systems: Networks of Plausible Inference*, Morgan Kaufmann, San Francisco, CA, 1988.

[PRO 59] PROGRAMS WITH COMMON SENSE, "Programs with common sense", in *Proceedings of the Teddington Conference on the Mechanization of Thought Processes*, Her Majesty's Stationery Office, London, 1959.

[RAO 95] RAO M., GEORGEFF P., "BDI-agents: From theory to practice", *Proceedings of the First International Conference on Multiagent Systems (ICMAS'95)*, Australian Artificial Intelligence Institute, Melbourne, 1995.

[SAE 17] SAE INTERNATIONAL, Automatic Emergency Braking (AEB) System Performance Testing, SAE Recommended Practice for Automatic Emergency Braking (AEB) system performance testing, J3087_201710, 2017.

[SOW 00] SOWA, J.F., *Knowledge Representation: Logical, Philosophical, and Computational Foundations*, Brooks/Cole, Pacific Grove, CA, 2000.

[SUT 98] SUTTON R., BARTO A., *Reinforcement Learning: An Introduction*, MIT Press, Cambridge, MA, 1998.

[THE 19] THEOCHARIS E., PAPOUTSIDAKIS M., DROSOS C. *et al.*, "Safety standards in industrial applications: A requirement for fail-safe systems", *International Journal of Computer Applications*, vol. 178, no. 24, pp. 0975–8887, 2019.

[VAL 84] VALIANT L.G., "A theory of the learnable", *Communications of the ACM*, vol. 2, no. 11, 1984.

[WEI 99] WEISS G., *A Modern Approach to Distributed Artificial Intelligence, Multiagent Systems*, MIT Press, Cambridge, MA, 1999.

[WIN 12] WINNER H., "Adaptive cruise control", in ESKANDARIAN A. (ed.), *Handbook of Intelligent Vehicles*, Springer, Cham, 2012.

[YAH 03] YAHIAOUI G., DA SILVA DIAS P., "On board visibility evaluation for car safety applications: A human vision modeling-based approach", *ITS Conference*, Madrid, 2003.

[ZAD 75] ZADEH L.A., "Fuzzy logic and approximate reasoning", *Synthese*, vol. 30, pp. 407–428, University of California, Berkeley, CA, 1975.

Conventional Vision or Not: A Selection of Low-level Algorithms

2.1. Introduction

In this chapter, the perspective adopted relates to a type of perception which resolutely privileges visual sensors – be they conventional or not – due to their low cost and the wealth of data resulting from them. The goal is to show how the relevant exploitation of their specificities (from the point of view of geometry and/or photometry), makes it possible to design low-level algorithms based on a "model", with characteristics as close as possible to such sensors, thus making it possible to draw a maximum benefit from their advantages. The first section will be dedicated to the presentation of the different types of visual sensors adapted to autonomous vehicle-related problems, and the following sections will discuss the algorithms that exploit the direct outputs of these sensors. These algorithms are all derived from a more or less dense relevant information set, retrieved from the images and their spatial and/or temporal correspondence. This type of information, called primitives, is presented in section 2.3.1. The following sections will be dedicated to the presentation of a choice of techniques, making it possible to show how low-level approaches integrate the notion of objectives as early as possible. In the field of autonomous systems, goals require high performance levels: detecting the navigable space, detecting obstacles (the vulnerable, two-wheeled vehicles, cyclists, other vehicles) or helping to estimate the ego-vehicle's own movement. These three examples of goals to be achieved can guide vision algorithms from the lowest level.

We will introduce a technique for detecting the navigable space, which has the particularity of relying on a method for accurately and densely estimating the optical

Chapter written by Fabien Bonardi, Samia Bouchafa, Hicham Hadj-Abdelkader and Désiré Sidibé.

flow. This method overrides the homogeneous road features that defeat the majority of motion estimation techniques. We will show how this technique is robust *vis-à-vis* the various disturbances that can affect road scenes and how it succinctly provides an estimate of 3D translations, suggesting a complete visual odometry technique as a perspective. We will then explain how accurate obstacle detection can be envisioned via ego-movement compensation and an estimation of the road scene flow. Finally, we will present an example of visual odometry using unconventional geometry.

2.2. Vision sensors

The purpose of Advanced Driving Assistance Systems (ADAS) is to provide the driver with safe tools that make the best use of embedded intelligence, based on both proprioceptive and exteroceptive sensors. During the last decade, many systems coupling a given set of sensors with adapted software solutions have been proposed. It should be noted that the question of the use of proprioceptive sensors no longer arises, the use of LIDAR sensors has become particularly widespread because it makes it possible to generate a high-resolution 3D maps of the environment, and visual sensors are often integrated in stereovision mode (thus avoiding scale estimation problems). These configurations, widely chosen in existing systems, raise the still open question about sensor cost and their operation in adverse or degraded environmental conditions (rain, fog, night, reflecting surfaces, glare, etc.). The answer to the first question invites us to pay special attention to visual sensors known for their low cost, the wealth of information they can provide and the ease for representing data in systems which can be understood by the human, in a context where approaches tend to be more "driver" centered. The second question concerning operation in degraded and adverse situations led researchers to become interested in both unconventional sensors and fusion approaches, making it possible to exploit the complementarity and/or redundancies of the different sensors, in order to provide more certain, reliable and robust measurements.

Vision-based sensors are generally the preferred choice of sensors to be implemented when instrumenting a vehicle. As they are affordable, they make it possible to acquire a large quantity and plurality of information that can be used for the perception of the environment. This non-negligible amount of information nevertheless requires a significant processing capacity, which can be critical when sizing an embedded system, such as a vehicle. The nature of the information useful for the perception and navigation of the vehicle that can be retrieved from images is rich: obstacle detection, estimation of the navigable space, recognition of signaling elements such as signs or traffic lights, detection of horizontal primitives (marking on the ground or on the sidewalk), staying in lane at medium and high speed, or parking assistance. The developments related to the autonomous vehicle require the transfer of methods used in the field of robotics, and employing visual sensors. The applications to be transferred concern the estimation of the vehicle's location and attitude, as well as the reconstruction of its overall trajectory, with the correction of system drifts.

2.2.1. *Conventional cameras*

So-called conventional cameras are the most common, and are often utilized for general public use. They are derived from the initial *Camera Obscura,* idea, a "dark optical chamber" pierced with a simple hole enabling the projection of a scene on a plane, also called the *pinhole camera.* The addition of lenses or objective systems has made it possible to concentrate more light inside the chamber, or to modulate the field of view (FOV) of the observed environment. Initially used for observation purposes (perception of perspective and its projections), systems have evolved to freeze images on photosensitive devices, such as metal plates with silver salts, as in the case of the *daguerreotype.* The rise of photography and digital sensors has replaced photochemical films with electro-photosensitive surfaces, matrices made up of photosites (namely, the *pixels* of a digital image), which generate an electrical signal depending on the amount of light received on their surface.

2.2.1.1. *The optical model of conventional cameras*

The projection model applied for conventional cameras is called the *pinhole model* or *pinhole* and is detailed in Hartley and Zisserman (2003). Represented in the diagram below, it models a perspective camera in a Euclidean space of dimension 3, by an image plane corresponding to the photosensitive sensor, as well as a center of projection, written as "C" hereafter. This model also defines an optical axis, that is to say, a straight line orthogonal to the image plane and passing through the center of projection **C**. We define the main point **p** as the intersection between the previously defined optical axis and the image plane. The distance between the center of projection **C** and the main point **p** is defined as the focal length, and depends on the optical device (lenses, objectives) fitted to the camera.

The pinhole model therefore expresses the relationship between the R_c points of the camera frame (projective space of dimension 3) at the R_i image frame (Euclidean space of dimension 2). Let $(x, y, 1)^T$ and $(X_c, Y_c, Z_c, 1)^T$ be the coordinates of the point projected on the image coordinate system and those of the point on the camera frame, respectively. We have the following relation with f, the focal length:

$$\begin{pmatrix} x \\ y \\ 1 \end{pmatrix} = P \begin{pmatrix} X_C \\ Y_C \\ Z_C \\ 1 \end{pmatrix} = \begin{pmatrix} f & 0 & 0 & 0 \\ 0 & f & 0 & 0 \\ 0 & 0 & 1 & 0 \end{pmatrix} \begin{pmatrix} X_C \\ Y_C \\ Z_C \\ 1 \end{pmatrix}$$

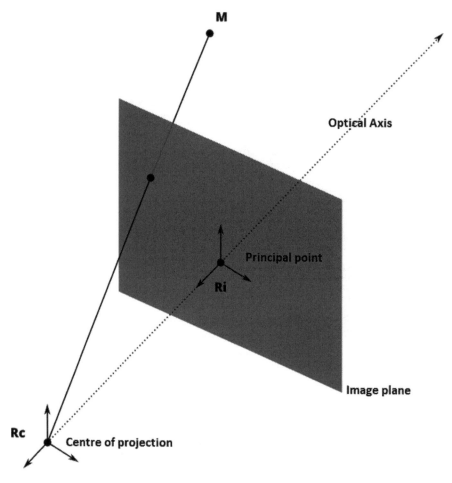

Figure 2.1. *Pinhole model. For a color version of this figure, see www.iste.co.uk/bensrhair/adas.zip*

We then define an affine transformation linking the coordinates in the image frame to the pixel coordinates $(u, v, 1)^T$ specific to the resolution, and to the placement of the photosensitive sensor on the image plane of the camera:

$$\begin{pmatrix} u \\ v \\ 1 \end{pmatrix} = A \begin{pmatrix} x \\ y \\ 1 \end{pmatrix} = \begin{pmatrix} k_u & -k_u/\cos\theta & u_0 \\ 0 & k_v/\sin\theta & v_0 \\ 0 & 0 & 1 \end{pmatrix} \begin{pmatrix} x \\ y \\ 1 \end{pmatrix}$$

where k_u and k_v are the number of *pixels* per unit of length along the two axes ($k_u = k_v$ if the pixels are square) and θ is the angle between successive lines of pixels. u_0 and v_0 represent the pixel coordinates of the principal point. We generally consider a simplified model by using an unbiased photosensitive matrix, with matrix A then being expressed as follows:

$$A = \begin{pmatrix} k_u & 0 & u_0 \\ 0 & k_v & v_0 \\ 0 & 0 & 1 \end{pmatrix}$$

It is also customary to define a matrix $K = AP$, which we call intrinsic parameters of the camera:

$$K = AP = \begin{pmatrix} f_u & 0 & u_0 & 0 \\ 0 & f_v & v_0 & 0 \\ 0 & 0 & 1 & 0 \end{pmatrix}$$

where $f_u = fk_u$ and $f_v = fk_v$.

The optical devices of conventional cameras have more or less pronounced defects due to the quality of lenses, for example, the optical properties of their materials, or their positioning. Thus, we can come across spherical and chromatic aberrations, a vignetting effect or even distortions. It is possible to model radial and tangential distortions, in order to process the images from the sensor, before applying the pinhole model when retrieving 3D information from the images.

2.2.1.2. *Spectral sensitivity of conventional cameras*

The spectral sensitivity of cameras varies from system to system. The most common technologies for conventional cameras are CCD (*Charge-Coupled Device*) and CMOS (*Complementary Metal-Oxide-Semiconductor*) sensors, which are sensitive to visible light, as well as to a part of the infrared spectrum. The sensors of consumer cameras are therefore mostly equipped with a physical infrared filter on their surface.

In order to transcribe the colors observed by the human eye, an additional filter associates each photosite with a particular color, green, red or blue. The distribution pattern of these colors often follows this matrix, although other patterns exist.

To obtain a final image composed of three color channels (RGB), an additional demosaicing or debayering stage (if it is a Bayer matrix) makes it possible to interpolate the adjacent photosites of different colors, so as to deduce the superimposed images specific for each channel.

2.2.2. *Emerging sensors*

2.2.2.1. *Unconventional geometry*

Many state-of-the-art optical devices have been proposed in order to extend the field of perception of a perspective (or conventional) camera. In particular, we can mention omnidirectional cameras, which enable the widening of the field of view in three different ways: by the use of a wide FishEye-type angle lens (dioptric systems), by the combination of a conventional camera and a catadioptric cone mirror system, and by assembling several cameras with different points of view (polydiopter systems). The so-called FishEye and catadioptric cameras are widely used for robotics and autonomous navigation. From a practical and theoretical point of view, central projection is a desirable property for establishing the camera's projection model. Indeed, a projection is said to be central if it complies with the single-viewpoint constraint, that is to say, all of the light rays converge towards a single point. In Baker *et al.* (1999) the authors describe the different single-viewpoint catadioptric solutions (central projection). Central catadioptric systems are generally the result of a combination of hyperbolic mirrors and a perspective camera, or parabolic mirrors and an orthographic camera. However, FishEye systems do not have the single-viewpoint property. In Courbon *et al.* (2007), it was shown that the unified central projection model initially described by Geyer *et al.* (2000a), and recalled below, can be used to model the strong distortions of a FishEye lens.

Unified central projection model: while projection models adapted to each configuration have been developed and presented in literature, it is possible to represent all types of single-viewpoint cameras using a unified central projection model, as proposed by Geyer and Daniilidis (2000a), for example.

Indeed, a central projection system can be modeled by two consecutive projections: a spherical projection followed by a perspective projection. This modeling has been widely exploited by the vision and robotics communities: projective geometry (Geyer *et al.* 2003; Hadj-Abdelkader *et al.* 2005), visual monitoring (Mei *et al.* 2008; Hadj-Abdelkader *et al.* 2012), calibration (Mei *et al.* 2007), visual servoing (Barreto *et al.* 2003; Hadj-Abdelkader *et al.* 2008a), for example.

Concretely, a 3D point of the scene defined by the coordinate vector $P = [X, Y, Z]^T$, is projected onto the unit sphere at a point of coordinates $P_s = [X_S, Y_S, Z_S]^T = \frac{1}{\rho}[X, Y, Z]^T$, with $\rho = \|P\| = \sqrt{X^2 + Y^2 + Z^2}$. The spherical point P_s is then projected onto the image plane at a normalized image point $m = [x, y]^T$ via a perspective projection, whose center of projection of the camera is at a distance ξ from the center of the sphere (see Figure 2.2), such that:

$$m = \begin{bmatrix} x \\ y \end{bmatrix} = \frac{1}{Z + \xi\rho} \begin{bmatrix} X \\ Y \end{bmatrix}.$$

The parameter ξ defines the type of camera represented by the unified model. In particular, it can be observed that when $\xi = 0$, the pinhole model is obtained.

In the end, the pixel image point is given by $p = Km$, where K is the calibration matrix containing the intrinsic parameters of the camera. Let us note that the matrix K and the parameter ξ can be obtained after calibration, using the method described in Mei *et al.* (2007), for example.

When the omnidirectional camera is calibrated, the spherical point P_S corresponding to a normalized image point, m can be calculated by inverting the unified projection model described previously. We therefore obtain:

$$P_S = \lambda [x \ y \ 1 - \xi/\lambda]^T$$

where $\lambda = \frac{\xi + \sqrt{1 + (1 - \xi^2)(x^2 + y^2)}}{x^2 + y^2 + 1}$

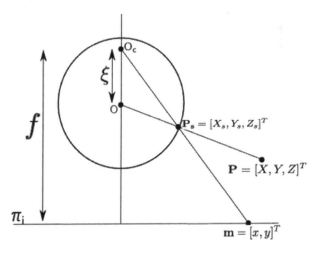

Figure 2.2. *Unified projection model*

2.2.2.2. *Unconventional photometry*

2.2.2.2.1. Multimodality and spectral bands

Visible cameras (colors or shades of gray) are vision-based sensors whose spectral response is identical or close to that of the human eye. However, there are matrix sensors of different spectral sensitivities that have some successful applications in aerial imagery (in meteorology or agronomy, for example), or even in medical imagery. The uses in the vehicle sector have successfully relied on infrared imagery for the detection of pedestrians or animals crossing the traffic lane, but the use of this type of sensor for localization tasks is still marginal. For example, Magnabosco *et al.* (2013) use a multimodal image registration method within a SLAM method. Maddern *et al.* (2012) also propose a method associating visible and thermal infrared images, and present both night and day recognition results. However, these approaches use each type of sensor independently: the problem of associating and comparing data from different spectral modalities remains an open question. In recent years, research has been carried out on the question of multimodality in robotics, or at least on the association of images of natural scenes. Some studies on the subject have been carried out by Ricaurte *et al.* (2014), for example. Others propose modifications to be made to local characteristic extractors, so as to make them invariant to modality changes (Firmenichy *et al.* 2011; Mouats *et al.* 2013).

2.2.2.2.2. Polarimetry

Recent vehicle-related research makes use of polarimetry to process or retrieve information from the observed environment. Over the past decade, this use has started to develop further, as a result of the marketing of compact and affordable polarimetric cameras, which can be mounted on vehicles, robots or drones (Rastgoo *et al.* 2018; Blanchon *et al.* 2019). There are different polarimetric sensor technologies, but we can also mention cameras based on DoFP (Division of Focal Plane), which have the advantage of enabling real-time acquisitions, while being robust. These cameras include a conventional sensor in the front, in which a micro-polarizer filter is placed, generally oriented at angles 0°, 45°, 90° and 135°. These four images make it possible to calculate the intensity, the degree and the angle of polarization at each point (Nordin *et al.* 1999; Guo and Brady 2000). In particular, these cameras make it possible to segment reflective or semi-reflective surfaces, such as glass windows, in order to infer the presence of vehicles (for example Fan *et al.* (2018); Blanchon *et al.* (2019)), or to eliminate interest points that may be observed through a reflection (reflective walls or wet roads) and thus become sources of error during the triangulation for a three-dimensional reconstruction of the environment.

2.2.2.2.3. Event cameras

So-called event or neuromorphic cameras are currently arousing genuine enthusiasm in the scientific community, particularly among roboticists. The best known, such as DVS (Dynamic Vision Sensor) (Lichtsteiner *et al.* 2008), aim to overcome the well-known drawbacks of conventional cameras, namely: temporal redundancy due to an arbitrary acquisition frequency, which does not depend on the changes/movements in the scene; a low dynamic range (around 60 to 70 db against 140 db for a natural scene); and a signal to noise ratio, strongly dependent on the lighting of the scene.

Event cameras are bio-inspired sensors that operate at the lowest level, in a radically different way from traditional cameras. Instead of acquiring images with a fixed acquisition frequency, the changes in brightness for each pixel are calculated in a completely asynchronous manner. This results in a flow of events, with the possibility of finding a *timestamp* for each event, its x and y position, and the sign indicating a change in brightness at that point, called polarity. Compared to traditional cameras, these have exceptional properties: a very high dynamic range (140 dB against 60 dB), a high temporal resolution (of the order of μs) and a very low power consumption. In addition, the movement of the objects or of the camera does not generate any blur. These advantages make them cameras with great potential for robotics and computer vision in scenarios that challenge traditional cameras, especially in the case of high speed and/or high dynamic ranges. However, their use requires redesigning and re-adapting classic algorithms. In recent years, there have been many contributions to dynamic vision in the fields of motion estimation, stereovision, visual odometry and localization (Gallego *et al.* 2019).

2.3. Vision algorithms

This section focuses on a selection of monocular or non-monocular vision algorithms, based on conventional sensors (or not), dense or non-dense primitives and those making it possible to exploit the geometric or photometric models of image formation. The algorithms presented show that it is possible to reach very high level objectives at a lower cost by exploiting the most basic information that can be retrieved from the images. These "low-level" approaches have the advantage of yielding interesting performances, while leaving space and a possible field of development for higher-level processes, which can only contribute to the improvement of the performances obtained.

2.3.1. *Choosing the type of information to be retrieved from the images*

Different types of information can be extracted (colors, shapes, contrasts or textures, for example), which may or may not be suitable for the intended application. We therefore choose particular characteristics or primitives called *features*. For an image considered as a whole without an *a priori* on the nature of the observed objects/scene, two approaches are possible: the first one is to extract the characteristics according to a fixed "grid", the second one involves detecting interest points.

Extraction following a fixed grid: a simple method to retrieve information from an image as a whole is to consider dividing the image into arbitrary zones. These areas can be squares of fixed dimensions (Sünderhauf *et al.* 2013; Naseer *et al.* 2014), but also other shapes for specific cases (when using Fisheye or omnidirectional optics, for example).

Detection of interest points: the notion of the interest point, introduced by Moravec (1977), makes it possible to characterize the areas of the image where the signal is rich in information. Thus, an interest point is a point in the image where the light intensity varies significantly in several directions (at least two directions simultaneously). The signal therefore contains more information at these points than at points corresponding to a one-dimensional change, such as the contour points. Much work has been done on the detection of interest points and one of the fundamental approaches, which has inspired many others, is the Harris and Stephen detector (Harris and Stephen 1988). This is based on the autocorrelation function of the signal, which will be briefly detailed below.

A measurement of local variations in the image I at point $x = (x, y)^T$, associated with displacement $\Delta x = (\Delta x, \Delta y)^T$, is provided by the autocorrelation function calculated on a W window centered at the point x through:

$$\chi(x) = \sum_{x \in W}[I(x) - I(x + \Delta x)]^2.$$

A first order approximation gives: $I(x + \Delta x) \simeq I(x) + \left(\frac{\partial I(x)}{\partial x} \frac{\partial I(x)}{\partial y}\right) . \Delta x$, which makes it possible to write the autocorrelation function as follows:

$$\chi(x) = \sum_{x \in W}\left[\left(\frac{\partial I(x)}{\partial x} \frac{\partial I(x)}{\partial y}\right) . \Delta x\right]^2 = \Delta x^T M(x)\Delta x,$$

where $M(x)$ is the autocorrelation matrix which represents the local variations of the image I at point x.

Point $x = (x, y)^T$ is considered an interest point, if for any displacement Δx, the quantity $\chi(x)$ is large. In other words, interest points are the points x for which the autocorrelation matrix $M(x)$ admits *two large eigenvalues*. To avoid the calculation of eigenvalues, Harris and Stephen (1988) propose the following operator:

$$k_H = \det(M) - \alpha trace(M)^2.$$

Interest points are obtained by taking the local maxima of this operator. Let us note that α is an empirically determined constant. Harris and Stephen suggest the value $\alpha = 0{,}04$.

An algorithm for detecting interest points performs the localization of the *feature* on the image (and can also determine its size, orientation or even its shape, in the case of advanced *blob* detectors). An extraction method also includes the phase which describes the *feature* (that is to say, the representation, by means of a vector, of the features locally extracted around the *feature*).

Local description: the vast majority of applications subsequently require the comparison and association of the different interest points retrieved from several images. We thus consider a region of interest around the detected point for which we calculate a descriptor. Moreover, in order to obtain an invariant description of certain geometric transformations, such as rotations and affine or projective transformations, this region of interest may require a sub-pixel interpolation stage calculating the descriptor. The term *local feature* designates an interest point accompanied by a region of interest chosen in its close neighborhood, and for which a description is calculated. The local feature represents a pattern of one or more pixels different from its close neighborhood. The quantified differences consider different properties depending on the algorithms: pixel intensity, colors or texture on a set of pixels. The search for local characteristics can be applied to a raw image in gray levels, or to a binary image resulting from an edge detection method applied upstream. The description is usually extracted according to a region focusing on the local characteristic. For some applications, it is then possible to directly match semantic information with the characteristic retrieved. Other approaches do not associate meaning with local features, but consider the accuracy of their localization and their distinctiveness from others *features* on the image. If their detection is stable over time, these local characteristics make it possible to calculate relative poses and calibrate the camera used, track observed objects or even perform a sparse three-dimensional reconstruction of the environment. These pose estimates and tracking methods make it possible to align and compose images, particularly while working with a close view field. One of the first detectors dedicated to this task is the *tracker* KLT (Lucas and Kanade 1981). A set of local features can also be used as the robust and compact representation of an image in its entirety. This makes it possible to recognize classes of objects, or even associate images representing the same place (we will then speak of "visual localization").

2.3.1.1. Properties of an ideal characteristic

Good quality *features* require the following properties (Tuytelaars *et al.* 2008):

– repeatability: a *feature* detected in one image should be detectable in another image, despite changes in scene viewing conditions (such as lighting);

– discriminant character: the regions described by the *features* must be sufficiently distinct, in order to limit confusion between two different patterns;

– locality: *features* must be "local" as much as possible, that is to say, they must represent a restricted set of pixels in order to reduce occlusion risks and ease the difficulties when estimating geometric transformations between two images;

– quantity: the bigger the number of extracted *features*, the greater the chance of describing a small object in the image. However, a large number of *features* can also lead to redundancies in the information retrieved;

– accuracy: a point must be detected with an accurate localization on the image for later use (mainly necessary for a calibration and a 3D reconstruction of the environment, as accurately as possible);

– performance: the processing time required for retrieving *features* can be decisive, particularly for embedded implementations.

2.3.1.2. Description methods

Global descriptors: global descriptors are extracted from images in their entirety (Chapoulie *et al.* 2011; Neubert *et al.* 2013; Naseer *et al.* 2014), or on medium sized patches (Lategahn *et al.* 2013). We thus calculate likelihood functions, such as *self-similarity* or mutual information. There are other possible approaches, such as GIST (Oliva *et al.* 2001), the application of Gabor filters or even wavelet representation, which is a form of frequency representation of the image's spatial information.

Interest point descriptors: the *feature* detection stage retrieves coordinates from the image which correspond to information source points (salient points, strong contrasts, etc.). Certain detectors also retrieve additional information, such as the size of the associated region of interest around the detected point, and possibly an orientation (this is the case of the SIFT detector, in particular). Depending on the detector chosen, after processing the image, patches of variable sizes and orientations are obtained. A descriptor represents the information of the interest points and their respective regions in the form of a vector of scalars, or a string of bits (binary descriptors). It is necessary to perform an interpolation of the regions to be described when they are not of the same dimension/orientation, in order to process identical resolutions. Scalar descriptors are a set of real values which correspond to gradients or to a spatial frequency representation, for example. Later, in order to determine whether they represent the same point or not, these vectors are compared using common mathematical distance operators. One of the most common scalar descriptors is SIFT (Lowe 2004), which is popular for its efficiency. The comparison of many points described by scalar vectors is highly demanding in terms of computing resources. Binary descriptors have been proposed to tackle this problem. Their output is a string of bits that can be compared using Hamming distance. Therefore, pairing calculations are much faster. However, these descriptors encode less information and are therefore more conducive to confusion between two different interest points.

Mid-level features and *pooling*: the usual detectors generally retrieve a high number of interest points: from a few hundred to several thousand, depending on the size of the image. The purpose of *Mid-level features* is to convert the set of descriptors corresponding to an object or scene observed, into a compact form. Many methods use *clustering* approaches to compute a dictionary. Every extracted descriptor is then associated with an element in this dictionary (for example, the closest element to the descriptor in the space of characteristics). Chatfield *et al.* (2011) establish a comparison of several techniques falling within the concept of *Mid-level features*. The figure below offers a representation of an approach using a dictionary. The *Bag-of-Words* approach can be classified among these techniques.

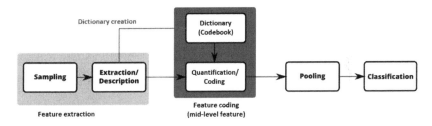

Figure 2.3. *General diagram of an approach using a dictionary. For a color version of this figure, see www.iste.co.uk/bensrhair/adas.zip*

Substitutions by neural networks (CNN): we can summarize the successive stages for comparing two images (query and reference) by means of the representation suggested in the figure below. Some tasks in the workflow can be replaced by a learning method. For example, Weyand *et al.* (2016), replace the entire processing chain by a CNN method.

Other authors use neural networks for more limited substitutions: in particular Aguilera *et al.* (2016) discuss the difficulty of associating patches extracted from different cross-spectral domains. These new methods are outperforming more traditional approaches. Learning is nonetheless very costly in terms of runtime and is only efficient for large training data sets.

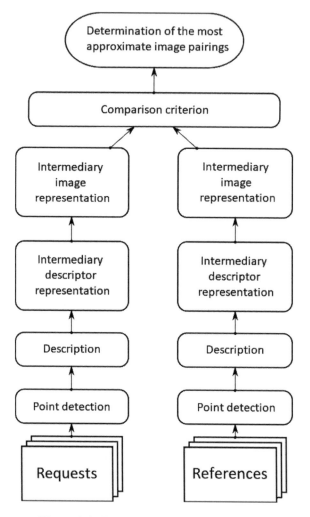

Figure 2.4. *Processing stages of a matching approach based on interest points*

2.3.1.3. *Interest-point matching process*

Matching processes involve identifying primitives or *features* in two or more "corresponding" images, for example, 2D primitives which are the projections of the same 3D point of the observed scene. There are two approaches: dense and sparse matching strategies. In the first approach, all visible pixels in the two images should be matched (non-occulted pixels), whereas in the second approach, matching concerns a certain number of particular primitives detected on the images, such as interest points.

Similarity measures: once the primitives have been detected and described by a feature vector, the matching process comes down to comparing the descriptor vectors. The choice of similarity measurement is therefore crucial. Correlation measurements and statistical distances, such as Mahalanobis distance or Bhattacharyya distance are commonly used.

Elimination of false matches: a verification step is necessary to eliminate incorrect pairings. An efficient approach is cross-checking (or cross-matching), which provides a symmetric set of pairs of interest points. Given two images I_1 and I_2, we start by matching the interest points of I_1 with those of I_2, then we exchange the roles of images I_1 and I_2. The pairs of correspondences finally retained are those formed by points which have been mutually selected, as shown in Figure 2.5. Another widely used approach relies on estimating the geometric transformation between two images, whenever this is possible. For example, it is possible to estimate homography or epipolar geometry using robust methods, such as RANSAC (Fischler and Bolles 1981).

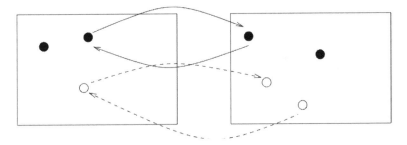

Figure 2.5. *Principle of cross-checking; the solid line indicates the correct correspondences; the dotted line, incorrect correspondences*

2.3.2. *Estimation of ego-movement and localization*

From the images acquired by the vision system embedded in the vehicle, several algorithms can be developed to perform various tasks, such as obstacle detection, estimation of the distance from an obstacle, recognition of object categories, localization of a vehicle in a given map, or SLAM (Simultaneous Localization and Mapping).

Each of these tasks and the description of the associated algorithms would require a full chapter, so in this section we will merely describe motion and ego-motion estimation algorithms: these are optical flow and visual odometry. These methods will be used in the two applications presented in the following sections.

2.3.2.1. *Estimation of the optical flow*

The optical flow – not to be confused with 2D movement, which is the actual projection of 3D movement on the image – is the apparent movement that is generated by the vehicle's ego-movement and the displacement of moving objects. The optical flow can be estimated directly from a sequence of images. This problem has been widely studied in literature, due to its many applications: detection and tracking of moving objects, autonomous navigation of robots and vehicles, visual odometry.

So-called differential methods are based on the calculation of spatio-temporal derivatives of the intensity function. They assume the hypothesis of brightness invariance during displacement, which amounts to considering that a point which moves along vector **u** keeps a constant brightness between two consecutive images at moments t and $t + dt$:

$$I(\mathbf{p}, t) = I(\mathbf{p} + \mathbf{u}, t + dt)$$

The approaches differ by the way in which this equation is developed, and the nature of the additional constraints used for estimating the speed vector at each point. The first-order Taylor's approximation has given rise to a class of so-called first-order differential methods. An equation known as the "motion constraint equation" is then obtained:

$$\frac{\partial I}{\partial x}u + \frac{\partial I}{\partial y}v + \frac{\partial I}{\partial t} = I_x u + I_y v + I_t = 0$$

The insufficiency of the motion constraint equation to estimate vector u at each point has led existing approaches to suggest additional hypotheses and/or constraints. In specialized literature, the methods are classified into two categories: those which provide local estimates based on the hypothesis that the optical flow is constant along small neighborhoods (Lucas and Kanade 1981), and global methods based on additional constraints which make it possible to provide dense estimates (Horn and Schunk 1981) via a process of regularization and propagation.

– Global methods: these approaches minimize an energy function made up of a data fidelity term, which reflects the invariance of brightness during displacement, and a regularization term, which expresses the slow variation of speeds between neighboring points (smoothing constraint):

$$\arg \min E = \int_{\Omega} [Ix(\mathbf{p})u(\mathbf{p}) + Iy(\mathbf{p})v(\mathbf{p}) + It]^2 + \lambda(||\nabla u(\mathbf{p})||^2 + ||\nabla v(\mathbf{p})||^2)d\mathbf{p}$$

Resolution methods are numerous and lead to iterative processes which hopefully lead to error convergence, by means of a prior estimation of the spatio-temporal gradients.

– Local methods: these approaches assume the initial hypothesis that vectors u are constant at an N given neighborhood:

$$\arg \min E = \sum_{k \in N \in (\mathbf{p})} (I_t(\mathbf{k}) + \nabla I^{\top}(\mathbf{k})\mathbf{u}(\mathbf{p}))^2$$

The optimal solution is obtained by solving the linear system $A\mathbf{u}=b$

with:

$$\mathbf{A} = \Sigma_{k \in N \in (p)} \nabla I^T(\mathbf{k})\nabla I(\mathbf{k}) \text{ and } \mathbf{b} = -\Sigma_{k \in N \in (p)} I_t(\mathbf{k})\nabla I(\mathbf{k})$$

2.3.2.2. Visual odometry and SLAM

The term "odometry" is used for estimating a trajectory from a set of relative position estimates. Visual odometry therefore involves estimating the trajectory of a moving vehicle incrementally from the images acquired during the displacement (Scaramuzza and Fraundorfer 2011). However, this technique inevitably leads to an accumulation of errors over time and therefore, to a deviation between the vehicle's estimated and actual trajectory.

Odometry is thus appropriate for short periods of time, whereas for long-term localization, it is preferable to have a map of the environment in order to avoid drifting the trajectory's estimate. Solving both localization and mapping problems simultaneously is known as SLAM, which stands for "Simultaneous Localization and Mapping" (Durrant-Whyte and Bailey 2006). This problem poses many challenges:

– *Initialization*: at the beginning of the process, there is no map to locate the vehicle. The initialization stage can be facilitated by the use of a stereoscopic system, but can be more difficult when only one camera is used.

– *Loop closing*: the map obtained by a SLAM algorithm is used for estimating the drift accumulated during the displacement, when the vehicle returns to an already-mapped place. This requires an ability to identify and pair known (mapped) and perceived (real-time perception) primitives. The quality of this pairing is directly related to the wealth of visual information available.

– *Relocalization*: the estimate of the current position of the vehicle is based on previous positions and the correspondences established between the visual sensor data and the map of the environment. In cases where the hypotheses concerning the vehicle's movement are not verified, or when the correspondences cannot be established due to occlusions, for example, it is necessary to develop a procedure for relocating the vehicle on the map. This procedure can also be used for establishing the vehicle's initial position.

– *Robustness*: the matching process between the sensor data and the map necessarily leads to errors, especially in dynamic environments (moving objects, variable environments, etc.). It is therefore necessary for algorithms to integrate procedures which can help detect and reduce the impact of false matches.

– *Real time*: algorithms must obviously be able to provide the vehicle with real-time information about its position on the map. They must also be scalable for long-term and large-scale uses.

Odometry versus SLAM: the essential difference between odometry and SLAM lies in the goal pursued. The purpose of SLAM is to estimate a vehicle's positioning by recalibrating a previously constructed map containing primitives, as well as selective and representative information. Localization is performed by *matching* the primitives resulting from embedded perception and the primitives already present on the map. The error between perceived and known primitives makes it possible to correct the current position of the vehicle. The map is global if the primitives are positioned with GSP coordinates. On the other hand, visual odometry aims to incrementally estimate the vehicle's trajectory, layer after layer. A local map of the environment is used to ensure the consistency of the estimated trajectory, but the mapping of the environment is not a goal in itself, unlike in the case of SLAM. Visual odometry can therefore be considered as a building block within a complete visual SLAM algorithm, but can also be used independently (Scaramuzza and Fraundorfer 2011). In this section, we will focus on the description of the principles of a visual SLAM algorithm, bearing in mind that this also applies to visual odometry. Finally, it is important to remember that when using a single camera for SLAM, we can only estimate the trajectory and the map of the environment up to a scale factor. The use of another source of information, such as an inertial unit, or the knowledge of certain distances from the map, are required to solve this scaling problem.

There are two main families of methods to solve the problem of visual SLAM: feature-based methods and direct methods.

Feature-based methods: these methods are based on the detection and mapping of primitives, such as interest points (Mur-Artal *et al.* 2015) or lines (Eade and Drummond 2009; Lemaire and Lacroix 2007) detected on the images. All of these methods are based on the minimization of a reprojection error, which is a geometric measure and has good convergence properties. More specifically, given a 3D point of coordinates X_W in the coordinate system and a 2D interest point with coordinates x_C, the reprojection error is written as follows:

$$e_{proj} = x_C - \pi_m(R_{CW}X_W + p_W),$$

where the rotation $R_{CW} \in SO(3)$ and the translation p_W transform the 3D point X_W of the scene into a 2D point of the image via the perspective projection model π_m.

Optimizing the positions of a set $\mathcal{P}$ of points and a set $\mathcal{C}$ of cameras, by minimizing the reprojection error, is called beam or *bundle adjustment* (Triggs *et al.* 2000), and is the basis of the most efficient SLAM and odometry algorithms. The beam adjustment is formulated as follows:

$$\{X_W^j, R_{CW}^i, p_W^i | \forall j \in \mathcal{P}, \forall i \in \mathcal{C}\} = \text{argmin}_{X_W^j, R_{CW}^i, p_W^i} \Sigma_{i,j} \rho \left(\| x_i^j - \right.$$
$$\left. \pi_m(R_{CW}^i X_W^j + p_W^i) \|_{\Sigma_i^j}^2 \right),$$

where x_i^j is the interest point associated with the 3D point X_W^j in the camera i, Σ_i^j is the covariance matrix of the position of the interest point x_i^j in the camera image i, $\|.\|_\Sigma$ is the Mahalanob distance, and $\rho(.)$ is a robust weighting function, such as the Huber function, to decrease the influence of incorrect correspondences.

The resolution of this optimization problem provides the poses (rotations and translations) of the whole set $\mathcal{C}$ of cameras, as well as the positions of all of the points in $\mathcal{P}$.

The advantage of these methods is that geometric primitives are easy to extract and manipulate. But their main limitation is precisely related to the use of such primitives, because the absence of characteristic primitives (a lack of texture, for example, or a lack of vertical primitives in the environment), or motion blur leads to bad estimates. In addition, these methods produce scattered maps of the environment, which are not very useful for tasks other than localization.

Direct methods: direct methods exploit the information drawn from the image pixels directly, without detecting geometric primitives. This makes them more robust in the case of weakly textured road scenes, or in the presence of motion blur. A distinction is drawn between dense methods, which use all the pixels of the image (Newcombe *et al.* 2011), semi-dense methods, which are limited to pixels with a high gradient value (Engel *et al.* 2014), and sparse methods, which use a subset of the image pixels (Engel *et al.* 2018).

Direct methods are based on the estimation of the depth associated with each pixel from the camera set, using photometric error. More precisely, given the 2D coordinates of a point x with an estimated depth d in camera i, the photometric error e_{photo} when this pixel is observed in the camera j is defined as follows:

$$e_{photo} = I_i(x) - I_j(\pi_m(R_{CW}^j(R_{WC}^i\pi_m^{-1}(x,d) + p_C^i) + p_W^j)),$$

where π^{-1} is the inverse projection model which calculates the 3D position X_C of a 2D point x depending on its depth d and the camera's intrinsic parameters:

$$X_C = \pi_m^{-1}(x,d) = \begin{bmatrix} d\dfrac{u - c_x}{f_x} \\ d\dfrac{v - c_y}{f_y} \\ d \end{bmatrix}.$$

Let us remark that the photometric error assumes Lambertian surfaces and therefore, direct methods fail in the presence of specular surfaces. It is nonetheless possible to perform a photometric calibration, as done in Engel *et al.* (2018).

The main limitation of dense and semi-dense methods is their complexity and long runtime, which does not enable joint optimization as for *bundle adjustment*.

Nonlinear optimization: the optimization methods used for visual odometry and SLAM are non-linear methods because the errors to be minimized (reprojection error or photometric error) are non-linear functions. Optimization algorithms such as Gauss-Newton and, in particular, the Levenberg-Marquadt algorithm are the most widely used (Nocedal and Wright 2006).

2.3.3. Detection of the navigable space by a dense approach

In itself, the 2D image obtained from a visual sensor contains a certain amount of information about the perceived environment. When the purpose is to avoid obstacles which are visually discriminating (in the sense that they have visual signatures/attributes differentiating them from other "objects" in the environment), it is possible to build models for detecting them. Several approaches have been proposed, exploiting symmetry, texture, or color. These fairly effective approaches, intended for "vehicle" shaped obstacles, find it difficult to detect or "recognize" a pedestrian, since shape variability and deformations make the task more complex. From that, the adoption of techniques based on pattern recognition and classification naturally followed. Among the proposed approaches, let us quote: binary cascading classification techniques, based on the use of a – generally linear – classifier; the use of Support Vector Machines or the use of deep neural networks. The advantage of these methods is that they make it possible to work on large spaces. Let us also mention *boosting* techniques, which aggregate several weak classifiers into one strong classifier. The results obtained by the strong classifier are greater than those obtained by each weak classifier. Commonly used methods are derived from the AdaBoost method. The use of this classifier is based on the definition of a representation basis projecting the images where training and detection tasks will be performed. An extremely popular solution for pedestrian detection is the use of HOGs (Histograms of Oriented Gradients). Other descriptors have been proposed, such as the Joint Ranking of Granules, HAAR wavelet decomposition, or PCA (Principal Component Analysis). These detection methods are also based on an offline learning phase. This learning is performed by showing the classifier a set of positive and negative examples. The representativeness of these training examples will strongly affect the performance of the classifier.

Along with appearance-based approaches, those based on the estimation of structural information about potential obstacles rely on a second camera, which relies on a stereoscopic vision process and helps estimate the depth z of perceived objects. In certain works, especially when the stereoscopic vision system is perfectly calibrated and rectified, obstacles are considered as being easily detectable front-parallel planes. Labayrade *et al.* (2002) defined a space, called V-disparity, in which the fronto-parallel planes are transformed into lines. After this study, it has been suggested to follow the same principle, considering horizontal planes represented by straight lines in the U-Velocity space (Labayrade *et al.* 2002; Wang *et al.* 2014). Afterwards, these are extracted using a Hough transform. We will come back to this technique which has inspired the work presented in this section. Beyond the simple detection of fronto-parallel or horizontal planes, the construction of maps, or more precisely of occupancy grids, has been fairly successful (Vu *et al.* 2008; Nedevschi *et al.* 2009). One of the main advantages of this approach is that the collaboration between several sensors is immediate and is of the "blackboard" type (that is to say, it relies on natural accumulation for sharing). Indeed, the same occupancy map can be populated indifferently using LIDAR, RADAR or stereovision points. Generally, it is the collaboration between LIDAR and stereovision that is most considered. On the other hand, the LIDAR will provide detection hypotheses which will be later confirmed by the vision (Labayrade *et al.* 2005; Labayrade *et al.* 2007; Rodriguez *et al.* 2010). The problem of obstacle localization can also be tackled by its dual: the identification of the free space in front of the vehicle. In that case, the problem would no longer be seeking to avoid potential threats, but seeking to define the space in which it is possible for the ego-vehicle to maneuver (Soquet *et al.* 2007).

Alongside "structure"-based approaches, the idea of bringing together motion estimation and structural estimation is interesting. Work that actively involves these two approaches has continued since the early 2000s, that is to say, since the available computing power has made it possible to carry out these two processes simultaneously. Among the most significant works, let us quote those described in Heinrich (2002), centered on the exhibition of an image invariant, in this case the ratio of the norm of the optical flow on the distance to the sensor. In addition, the principle of 6D-Vision, presented in Franke *et al.* (2005) also constitutes an interesting approach. This is based on the monitoring of interest points using *Kalman* filters, tuned to the movements likely to animate the objects in the scene. As we have seen above, the formalism of the occupancy grids (probabilistic, credibilistic, fuzzy, and belief plot) favors an easy integration of different sensors. It is therefore natural to find its application here, in order to encourage the cooperation between artificial vision methods. This formalism can be exploited in order to construct a representation of the observed scene (Dornaika *et al.* 2000), or can simply be enriched with temporal information (Braillon *et al.* 2008; Leibe *et al.* 2007). The approaches receiving the greatest interest from the community are those centered on the evaluation of the *scene-flow* (Bak *et al.* 2010), or the extension of the optical flow to a three-dimensional space. In order to achieve this, it is possible to follow interest points (Lenz *et al.* 2011), or to integrate stereo into a method for calculating the optical flow, following the Horn & Schunk model (Pons *et al.* 2007; Wedel *et al.* 2008). From this correspondence field, classical segmentation techniques can be used to obtain a representation of the scene as a function of the apparent movement of the objects.

In the following two subsections, a cumulative monocular approach will be detailed. This approach makes it possible to use an estimated optical flow, as accurately as possible, in order to detect the navigable space. This is likened to a 3D, quasi-horizontal plane, whose properties in the associated 2D motion field are sufficiently distinctive so as to allow for on-the-fly detection.

The following sub-section focuses on the first stage: proposing a new method for estimating the optical flow from a sequence of images as accurately as possible. The approach is based on the generation of a quality map to refine the estimates via an iterative process, based on criteria of color similarity or proximity. The motion map thus obtained is then used for the rapid detection of the main 3D planes. For this, a cumulative approach, called *UV-velocity* has been developed. This approach exploits the geometric properties of the 2D velocity vector field. It makes it possible to detect planar surfaces using the hypotheses related to the nature of ego-movement as a starting point. The suggested method enables a progressive voting strategy which takes into account different ego-movement models and surfaces. The motion model for each detected surface is reintegrated into the optical flow estimation method, which then becomes an optimization method under the validity constraint of the planar model, thereby improving the accuracy of the optical flow. Furthermore, we show how a visual odometry process can benefit from the planar surface detection method. The optical flow estimation approach is evaluated in terms of accuracy and the execution time on the *Middlebury*[1] database. Regarding UV-*velocity*, validation is done on both simulated flows and on images from the KITTI[2] database.

2.3.3.1. *Optical flow estimation, confidence and reliability*

Regardless of the nature of the method used – local or global – the detection of the navigable road space by relying on an estimation of the optical flow is a risky bet. Navigable areas are often homogeneous. This makes the field of estimated speed vectors have erroneous orientations, mainly due to the classic problem of opening. The accuracy of the optical flow becomes important when the field properties of the vectors are exploited to densely estimate the location of the navigable area. It becomes crucial when vector analysis also allows for an on-the-fly estimation of the ego-movement.

The goal of this section is to show how a local classical approach to the optical flow can be adapted in view of improving the accuracy of estimations, by taking into account a *confidence* measure *a priori*, and a self-assessment measure *a posteriori*, which will be designated as *reliability*. We will show how confidence and reliability can be combined into a single *quality* score, on which a propagation process will be based, whose result will be a more accurate and dense optical flow. Starting from a classical optical flow estimation method such as KLT, we will introduce:

– a classic measure of confidence called *cornerness*, making it possible to filter the estimates likely to provide poorly conditioned matrices, within the framework of an approach derived from the KLT method;

– a reliability index capturing the temporal evolution of residuals during the convergence of KLT;

– a reliability index reflecting the local coherence of optical flow vectors in the neighborhood.

From these three measures assessing confidence and reliability, an overall quality map can be constructed. The optical flow is refined from a first raw estimate by using the quality map on an *a priori* set of starting seeds, chosen from the points

1 Available at: http://vision.middlebury.edu/flow/data/.

2 Available at: http://www.cvlibs.net/datasets/kitti/.

with the best quality score. The neighboring vectors are then corrected in two steps: a sparse correction of the seeds, followed by a dense correction on the entire map (Mai *et al.* 2017a; Mai *et al.* 2020).

2.3.3.1.1. Confidence measure

In literature, there are few existing approaches measuring the confidence of an estimate of optical flow. When they do exist, they are rarely exploited to improve estimates. The most textured areas of the image are often those providing the best estimates, due to the high informational content they possess. The most common criterion is the so-called *cornerness* measure, used in the *KLT* method suggested by Shi and Tomasi (1994). The areas of the image providing high confidence are those for which the two eigenvalues of the Hessian matrix are high:

$$\mathbf{J} = \begin{bmatrix} \sum I_x^2 & \sum I_x I_y \\ \sum I_x I_y & \sum I_y^2 \end{bmatrix} \rightarrow w_{corner} = \min{(\lambda_1, \lambda_2)}$$

This measurement proves to be more relevant than a simple indicator based on gradients, since it makes it possible to capture the structure and the local form of the intensity function in the zone considered.

2.3.3.1.2. Reliability measures

– Criterion for local uniformity of movement

A detailed analysis of the distribution of speed vectors in the space (u, v) makes it possible to study their distribution and their distance in relation to the constraint line: $I_x u + I_y v + I_t = 0$ where $\nabla I = (I_x, I_y, I_t)^T$, which they are supposed to verify because they are deduced from the brightness invariance equation. At the edge of different motions, vectors are dispersed and distant from the constraint line, thus leading to erroneous estimates. One way to measure the reliability of the estimates involves measuring the dispersion of the vectors over a given neighborhood. This is designated as S, the set of estimated flows around the point $\mathbf{p}$. The reliability S_{var} of the estimated vectors is calculated based on the variance $\sigma_S^2(\mathbf{p})$: $S_{var}(\mathbf{p}) = \frac{1}{\sigma_S^2(\mathbf{p}) + \varepsilon}$. By normalizing S_{var} between $[0,1]$, we obtain the reliability measure based on a local uniformity of the movement criterion: $w_{var}(\mathbf{p}) = \frac{S_{var}(\mathbf{p})}{\max{(S_{var})}}$.

– Criterion for the temporal evolution of residuals

As the KLT method is based on an iterative minimization method, in theory, convergence is ensured when the residue to be minimized decreases from one iteration to another, and gradually approaches 0 (zero) value. In practice, convergence may never occur, and an oscillation around zero may cause unstable behavior. The stability of the residue can be assessed by means of two criteria. The first compares the evolution of the residue from one iteration to another, in order to check its decrease:$w_{\Delta\epsilon} = max\left(0, \frac{\epsilon_k - \epsilon_{k+s}}{\epsilon_k}\right)$. The second calculates the maximum value of the residuals collected on the whole of the image, in order to normalize the value of the residue:$w_{S_\epsilon(x,y)} = \frac{max_{(x,y)}\left(S_{k,k+s}(x,y)\right) - S_{k,k+s}(x,y)}{max_{(x,y)}\left(S_{k,k+s}(x,y)\right)}$. The final temporal evolution criterion is a weighted sum calculated in accordance with the two previous criteria: $w_{res}(\mathbf{p}) = 0{,}5w_{\Delta\epsilon}(\mathbf{p}) + 0{,}5w_{S_\epsilon}(\mathbf{p})$.

2.3.3.1.3. Final quality score

The final quality score is the minimum between the confidence measure (which allows for a better confidence on the textured zones), the local uniformity of movement criterion (which offers a better consistency of the estimates) and the residuals temporal evolution criterion (which favors the stability of the estimates): $w_{mix} = min\left(w_{corner}, w_{res}, w_{var}\right)$.

2.3.3.1.4. Propagation process

The quality score is now used in a two-level propagation process: a sparse level and a dense level. At the sparse level, the points having collected the maximum quality scores are considered reliable enough to base the propagation on their own estimates. These are called "seeds". However, a first correction stage is carried out over these points:

– *Sparse optical flow correction*

For each selected seed, a refinement is made by considering the neighboring seeds on a given window. The flow is corrected by considering the degree of similarity (particularly the color attribute, if this is available) between the current seed and its neighbors:$\hat{u}(\mathbf{p}) = \frac{\sum_{i \in N(x), i \neq x} e_{simi}(\mathbf{p},i)u(i)}{\sum_{i \in N(x), i \neq x} e_{simi}(\mathbf{p},i)}$. The reliability score is simultaneously updated as follows: $\hat{w}(\mathbf{p}) = \frac{\sum_{i \in N(x), i \neq x} e_{simi}(\mathbf{p},i)w(i)}{\sum_{i \in N(x), i \neq x} e_{simi}(\mathbf{p},i)}$

with: $e_{simi}(\mathbf{p}, i) = e^{-\frac{d_{color}(\mathbf{p},i)}{\sigma_c} \frac{d_s(\mathbf{p},i)}{\sigma_s}}$

where $d_{color}(\mathbf{p}, \mathbf{i})$, $d_s(\mathbf{p}, \mathbf{i})$ are the Euclidean RGB distance between the colors and the distance between the points $\mathbf{p}$ and $\mathbf{i}$, respectively, and σ_c, σ_s are the decay rates of the color and spatial distances. The optical flow remains unchanged if its reliability is better than the calculated score: if$\hat{w}(\mathbf{p}) > w(\mathbf{x})$, then $u(\mathbf{p}) = \hat{\boldsymbol{u}}(\mathbf{p})$ and $w(\mathbf{p}) = \hat{w}(\mathbf{p})$.

– Dense optical flow correction

The correction method described above, which acts on the selected seeds, is then generalized to all points. This would therefore provide a dense propagation, guided by reliability scores.

2.3.3.1.5. Results

Our approach has been compared and integrated into the *MiddleBury* ranking, with a score of 86.9/158 in 2018. Bearing in mind the ground truth provided in *MiddleBury*, we chose to compare it with several approaches in terms of mean angular accuracy (where AE stands for *Angular Error*) and mean accuracy on vector end points (EP stands for *End-Point error*): Layer ++ relates to (Sun *et al.* 2010; Brox *et al.* 2004), LDOF (Brox *et al.* 2011) and FOLKI (Plyer *et al.* 2016). The results obtained in terms of accuracy show and confirm the relevance of a regularization based on confidence and quality indicators (Mai *et al.* 2020).

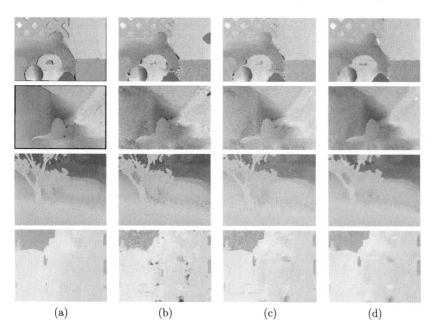

(a)   (b)   (c)   (d)

Figure 2.6. *Qualitative results on Middlebury's sequences (a) ground truth flow (b) results of the KLT method, (c) result of our approach before propagation (d) results obtained by our approach after propagation. For a color version of this figure, see www.iste.co.uk/bensrhair/adas.zip*

2.3.3.2. *Exploitation of the optical flow for the detection of 3D planes*

In this subsection, we consider that the optical flow is an approximation of the 2D velocity field on the image. The latter is the projection of 3D speeds on the image plane, according to a classic pinhole-type projection model. Our starting point is a precise estimation of the optical flow, as presented in the previous subsection, and we wish to show that an analysis of the geometric properties of the vector fields on the image will make it possible to define a voting strategy which provides the output both for the main planes of the scene (including the road) and for the translational movement of the embedded camera.

Let us consider the coordinate system $OXYZ$ fixed on the center of projection of a camera. The OZ axis coincides with the optical axis if we consider a rigid movement of the sensor, characterized by its instantaneous translational speed $\mathbf{T} = (T_X, T_Y, T_Z)$ and its instantaneous rotational speed $\Omega = (\Omega_X, \Omega_Y, \Omega_Z)$. Each point $\mathbf{P} = (X, Y, Z)$ belonging to the static scene has a relative movement $V = -T - \Omega \times \mathbf{P}$. If we consider that the projection of the point $\mathbf{P} = (X, Y, Z)$ according to the pinhole model on the image plane is $\mathbf{p} = (x, y, z)$ and that the focal length is f, then the 2D speed (u, v) at each point of the image is:

$$u = \frac{xy}{f}\Omega_X - \left(\frac{x^2}{f} + f\right)\Omega_Y + y\Omega_Z + \frac{xT_Z - fT_X}{Z}$$

$$v = -\frac{xy}{f}\Omega_Y + \left(\frac{y^2}{f} + f\right)\Omega_X + x\Omega_Z + \frac{yT_Z - fT_Y}{Z}$$

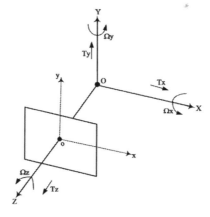

Figure 2.7. *Considered system and reference*

The examination of these equations suggests several remarks:

– 2D movement depends on depth;

– only the translational component of movement depends on depth;

– any 2D movement discontinuity can only be due to a variation in depth. Motion can only be determined up to a scale factor. Therefore, an object located at a distance Z, translating at T will produce the same 2D motion as an object at a distance kZ, translating at kT.

The dependence of the preceding equations on depth assumes that the latter is known upstream, or that the visual sensor is used in stereovision mode. As part of our study, we wish to take the possibilities of monocular vision to the maximum. One way to break free from the knowledge of depth at each point is to assume that the scene is made up of a set of planes (scenes following the "Manhattan world" model), in order to introduce a relationship between x, y and Z, thus making it possible to eliminate Z. Let us suppose that the camera observes a planar surface of the equation: $\mathbf{n}^T\mathbf{P} = d$, where $\mathbf{n} = (n_X, n_Y, n_Z)$ is the normal unit vector for the surface, d is the "plane/origin" distance and $\mathbf{P}$ is the coordinate point (X, Y, Z). Starting from equations [2.17] and [2.18] and from $\frac{1}{fd}|n_X x + n_Y y + n_Z f| = \frac{1}{Z}$, the 2D speed can be written out if we first place ourselves in the context of a pure translational movement (Longuet-Higgins and Prazdny 1980; Verri Poggio 1989), with f being the focal length of the system integrating the pixel/mm scale change:

$$u = \frac{xT_Z - fT_X}{fd}|n_X x + n_Y y + n_Z f|$$

$$v = \frac{yT_Z - fT_y}{fd}|n_X x + n_Y y + n_Z f|$$

To start with, let us consider the hypothesis that the vehicle's navigation requires at least a rough labeling of the road scene. Objects are labeled according to their nature and structure (horizontal surface = road; vertical surface = building; frontal surface = obstacle) and depending on their movement (movement in accordance with the vehicle's ego-movement = static object; movement not in accordance with the vehicle's ego-movement = objects with independent motion). For each of the surface models, Table 2.1 shows the expression of the associated u and v vectors. An analysis of this table lets us conclude that for horizontal planes, u is a function of x and y, whereas v is a quadratic function of y. For the lateral planes, u is a quadratic function of x, and v is a function of x and y. Finally, for the frontal planes, u is a linear function of x, and v is a linear function of y.

	U	V
Horizontal	$u = \dfrac{T_Z}{fd} x\lvert y\rvert - \dfrac{T_X}{d} \lvert y\rvert$	$v = \dfrac{T_Z}{fd} y\lvert y\rvert - \dfrac{T_Y}{fd} \lvert y\rvert$
Lateral	$u = \dfrac{T_Z}{fd} x\lvert x\rvert - \dfrac{T_X}{d} \lvert x\rvert$	$v = \dfrac{T_Z}{fd} \lvert x\rvert y - \dfrac{T_X}{d} \lvert x\rvert$
Frontal	$v = \dfrac{T_Z}{fd} x - \dfrac{T_X}{d}$	$v = \dfrac{T_Z}{fd} y - \dfrac{T_Y}{d}$

Table 2.1. *Surfaces motion in a so-called Manhattan world model*

The exploitation of the relations between the 2D speed and the curves defined on the image which only depend on x and y, is an old idea which has been little exploited for the detection of navigable surfaces in the context of ADAS applications. Two existing studies in specialized literature caught our attention:

– In the first study (Fermuller and Aloimonos 1995), the speed vectors of a given amplitude and orientation are constrained to belong to curves within the image whose parameters depend on the parameters of the sensor's 3D motion. In particular, if the vectors considered result from an optical flow or from a disparity field, then we can show that these are constrained to belong to conic sections, which can be determined. By studying the curves' properties, it is possible to draw an estimate of the ego-movement.

– In the second case (Labayrade *et al.* 2002), a stereovision study, the authors suggest a very effective technique based on the concept of *v-disparity*, which exploits the relationship between disparity and the lines of the image in the specific case where the stereoscopic images are rectified. A new cumulative projection space – the *v-disparity* space – calculated from the disparity histograms of all the lines on the image, makes it possible to highlight the proportionality ratio between the disparity and all of the image lines in the particular case of a horizontal plane.

These two studies can be approached in parallel in an interesting way: they both exploit iso-value curves: speed for one, disparity for the other. Our point of view is that this process, based on the definition of iso-curves on one hand and on the statistics along these curves, on the other hand, can be further exploited. That is why we naturally proposed its generalization to the case of monocular vision. In the approach that we develop, the estimation of the speed vectors by a dense optical flow method generates a population of voters of a substantial size, making the cumulative decision process representative and statistically relevant. For a given motion model and nature structure (an orientation plane located at a specified distance from the origin of the camera frame), "primitives" are associated with a speed. This is on an iso-speed curve *reinforcing the hypothesis of conjectured movement and structure.* The question that arises after analyzing Table 2.1 is that of the design of adequate voting spaces, which make it possible for the desired structures to be exhibited. The equations lead us to two spaces $U(u, x)$ and $V(v, y)$. Each point (x,y) of the image, associated with its 2D motion vector assimilated to the optical flow (u,v), respectively, vote for the localization of (u, x) and (v, y) in the two cumulative spaces **U** and **V**. Given the ratios presented in Table 2.1., the points belonging to horizontal planes (road) then reveal parabolas in **V**; points belonging to lateral planes (buildings in urban scenes) exhibit parabolas in **U**; whereas the points belonging to the frontal planes (obstacles) simultaneously exhibit straight lines in **U** and **V**. If we neglect the absolute values that we will consider later we then find that the curves formed in spaces **U** and **V** (which we will call *uv-velocity* spaces by analogy to *v-disparity* and *v-velocity*, a version of the algorithm using the speed vector moduli (Bouchafa *et al.* 2012a, 2012b)) are linear and follow the pattern: $u = Ax + B$ or $v = Ay + B$ (frontal plane), or parabolic and follow the pattern: $v = Ay^2 + By$ (horizontal plane) or $u = Ax^2 + Bx$ (side plane). Let us focus on a horizontal plane for which the relationship in v and y is parabolic, and analyze this space in order to find a way to make these parabolas appear effectively. Let us remember that in this specific case:

$$v = \frac{T_Z}{fd}y|y| - \frac{T_Y}{d}|y| = \frac{T_Z}{fd}(y|y| - |y|y_{FOE})$$

where we introduce the coordinates of the Focus of Expansion (FoE) (x_{FOE}, y_{FOE}), in order to center the image coordinate system on this point, whose coordinates are: $x_{FOE} = f\,T_X/T_Z$ and $y_{FOE} = f\,T_Y/T_Z$. All of the equations are therefore expressed in relation to this reference. Given $K = |T_Z|/fd$. the previous equation can then be rewritten by removing the absolute values:

$$v = K(y^2 - yy_{FOE})\mathrm{sign}(yT_Z)$$

This equation can be expressed in the following form:

$$v = Kv'\mathrm{sign}(yT_Z)$$

where $v' = (y^2 - yy_{FOE})$.

Let us note that v' is the equation of a parabola in the image, which is fully defined if the Focus of Expansion is known. The component v of the optical flow then varies linearly with v', and K is the constant associated with this linear relation. Knowing v by means of a given optical flow estimation method, and calculating v' from the knowledge of the position of the Focus of Expansion, constant K can be determined by a simple vote, making it possible to evidence the linear relationship between v and v' and constant K on-the-fly, which will lead to estimating the 3D translation parameters of the embedded camera. The same reasoning can be applied to the lateral planes. As far as the frontal planes are concerned (associated with the obstacles), the relationships are easier to handle because they are already linear: $v = \frac{T_Z}{fd}(y - y_{FOE}) = Ky'$ and $u = \frac{T_Z}{fd}(x - x_{FOE}) = Kx'$. Once the different curves have been detected in the U and V voting spaces by a simple *Hough* transform, a re-projection on the image makes it possible to highlight the horizontal, lateral and frontal planes having obtained the maximum vote.

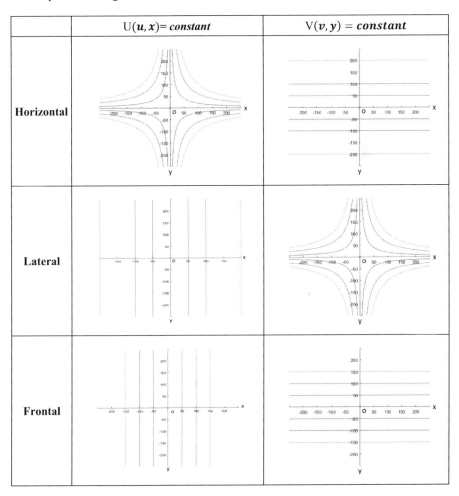

Table 2.2. *Iso-speed curves associated with each plane model in an urban scene (horizontal = road, lateral = building, frontal = obstacle). The cumulation is done along these curves on the image, in order to detect each plan category. For a color version of this table, see www.iste.co.uk/bensrhair/adas.zip*

The promising approach on simple motion models can be generalized, with no difficulty, to planes having a non-zero orientation, which can be categorized as *ex-horizontal* for $\mathbf{n} = (0, \sin{(\theta)}, \cos{(\theta)})^T$ and as *ex-vertical*, for those whose orientation is $\mathbf{n} = (\sin(\phi), 0, \cos(\phi))^T$. Likewise, the gradual introduction of 3D rotations generalizes the exercise and makes it possible to show that the cumulative approach is robust and that spaces $\mathbf{U}$ and $\mathbf{V}$ remain generic and usable for the detection of inclined planes.

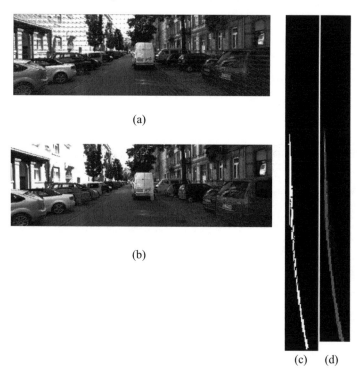

(a)

(b)

(c)   (d)

Figure 2.8. *(a) Estimated optical flow. (b) Result of segmentation using uv-velocity. (c) Cumulative space V (v, y). (d) Parabola detected after Hough transform. For a color version of this figure, see www.iste.co.uk/bensrhair/adas.zip*

The results presented above show how conventional monocular vision, without temporal integration, makes it possible to obtain the 3D plans of a scene on its own, as in "Structure From Motion" approaches, which led to the generalization of SLAM (Simultaneous Localization and Mapping) techniques. The relationship between structure and movement is such that knowing the structure already signifies a step forward towards estimating 3D movement, thus leading to what is commonly called visual odometry. Visual odometry is therefore the first step of a SLAM technique, which differentiates itself thanks to the addition of a detection and loop closing process, as well as incremental mapping. The following section will show how the previously introduced plane detection technique can help obtain a rough estimate, while some 3D translational motion parameters are operating. This estimated data can be used in visual odometry construction.

Figure 2.9. *Images showing the detection of the raw road (horizontal plane), without any spatial or temporal filtering. For a color version of this figure, see www.iste.co.uk/bensrhair/adas.zip*

Sequence	uv-velocity	Road plane detection using Homography
1	0.95	0.86
2	0.88	0.66
3	0.78	0.77

Table 2.3. *Comparison of the segmentation accuracies for uv-velocity and a conventional detection method, using homography on three of the sequences from Figure 2.9*

Motion	$(T_X; T_Y ; T_Z) \neq (0; 0; 0)$ $(\Omega_X; \Omega_Z) = (0; 0); \Omega_Y \neq 0$	$(T_X; T_Y ; T_Z) \neq (0; 0; 0)$ $(\Omega_Y; \Omega_Z) = (0; 0); \Omega_X \neq 0$
Equations	$$u = -\left(\frac{x^2}{f} + f\right)\Omega_Y + \frac{xT_Z - fT_X}{Z}$$ $$v = -\frac{xy}{f}\Omega_Y + \frac{yT_Z - fT_Y}{Z}$$	$$u = \frac{xy}{f}\Omega_X + \frac{xT_Z - fT_X}{Z}$$ $$v = \left(\frac{y^2}{f} + f\right)\Omega_X + \frac{yT_Z - fT_Y}{Z}$$
Ex-horizontal	$$u = -\left(\frac{x^2}{f} + f\right)\Omega_Y + \frac{xT_Z - fT_X}{fd} \times$$ $$\lvert ysin(\theta) + fcos(\theta)\rvert = F(x,y)$$ $$v = -\frac{xy}{f} + \frac{yT_Z - fT_Y}{fd} \times$$ $$\lvert ysin(\theta) + fcos(\theta)\rvert = F(x,y)$$	$$u = \frac{xy}{f}\Omega_X + \frac{xT_Z - fT_X}{fd} \times$$ $$\lvert ysin(\theta) + fcos(\theta)\rvert = F(x,y)$$ $$v = Ay^2 + By + C = \mathbf{F}(y)$$
Ex-lateral	$$u = Ax^2 + Bx + C = \mathbf{F}(x)$$ $$v = -\frac{xy}{f}\Omega_Y + \frac{yT_Z - fT_Y}{fd} \times$$ $$\lvert xsin(\phi) + fcos(\phi)\rvert = F(x,y)$$	$$u = \frac{xy}{f}\Omega_X + \frac{xT_Z - fT_X}{fd} \times$$ $$\lvert xsin(\phi) + fcos(\phi)\rvert = F(x,y)$$ $$v = \left(\frac{y^2}{f} + f\right)\Omega_X + \frac{yT_Z - fT_Y}{fd} \times$$ $$\lvert xsin(\phi) + fcos(\phi)\rvert = F(x,y)$$

Table 2.4. *Surface motion by introducing inclined planar surfaces and 3D rotations*

Figure 2.10. *Left: estimated optical flow. The sequences are associated with 3D movements, including rotations (turns). Right: detected ex-horizontal and ex-lateral planes. For a color version of this figure, see www.iste.co.uk/bensrhair/adas.zip*

2.3.4. *From the detection of 3D plans to visual odometry*

In the previous section we showed how, by knowing the FoE (Focus of Expansion), the 3D planes of the scene can be represented in the cumulative space using parabolas. Let us recall that the position of the FoE in the image is associated with the parameters of the sensor's 3D translational movement. In addition, the lower the error on the FoE estimate, the finer the obtained parabolas will be in the appropriate voting spaces. We then found it interesting to exploit this bias to estimate the actual FoE position, and therefore, the translational movement parameters, which in itself constitutes a step towards complete visual odometry. This sub-section will show how knowing a plane can conversely enable the detection of the FoE. For this, it is necessary to not only be able to isolate the different planes from the image, but also to quantify the parabolic aspect of the representations of these planes, by defining a metric that reflects the distance separating a presumed FoE from the real FoE. Let us detail the process below.

Definition of a suitable metric for the localization of the FoE: first, let us suppose that the image contains at least one plane. This plane, considered, for example, as a vertical plane, is written as π. The dispersion of the representation of a plane in the voting space – as defined in the previous section – can be formalized as the sum of the elementary widths of the curve representing π; these widths are considered as the squares of the deviations from the local mean. It therefore seems interesting to use this dispersion as a metric, monotonic with the distance separating the supposed FoE from the real FoE. In addition, it is convex and can enable the use of conventional optimization techniques, in order to find the position of the FoE. Starting from this measure, it is possible to formalize the FoE localization problem as a least-squares problem, which can be solved thanks to a classical optimization scheme (Bak *et al.* 2011a; Bouchafa *et al.* 2012a, 2012b), for example. The use of a gradient descent seems well suited. The approach in its first version requires a preliminary extraction of vertical and horizontal planes. In our first experiments, we opted for the extraction of horizontal, vertical and frontal planes.

Synthetic image results: first, the approach was tested on synthetic images. In addition, we performed a study of its sensitivity to several factors: the noise over the optical flow, the influence of the sensor's rotations, as well as the variation in the number of planes used. These first tests led to the following conclusions: (1) on a synthetic optical flow, and in the absence of noise, an exact extraction of the position of the FoE can be correctly carried out, whatever its position; (2) after adding various types of noise to the synthetic flow, it appears that the location of the FoE is very robust in relation to these disturbances; (3) large rotation rates (above 0.05 radian/frame) introduce a fairly large error. For weaker rotations, with the order of magnitude of those which can appear as a "straight line" during a trajectory, this error is much more limited; (4) we did not observe any remarkable influence regarding the number of plans used. It still is necessary to verify whether the nature of the plans (road, building, obstacle) plays an important role.

Pseudo-realistic image results: this method was tested on pseudo-realistic images generated by the Pro-SIVIC simulator, developed by IFSTTAR (now Gustave Eiffel University) and LIVIC. This platform is now marketed by ESI group. The simulator has synthesized a sequence of 250 pairs of stereo images in an urban environment with moderate traffic. In this sequence, the movement of the vehicle is mainly translational. Figure 2.11 shows an example of FoE extraction. It can be observed that the extracted FoE rests on the horizon line, something which is consistent with the known movement of the mobile. For this image, the error made between the extracted FoE and the real FoE – recalculated from the movement's components – is 2.2 pixels. Over the entire sequence considered, the average error made is 5.2 pixels, with a maximum error of 15.6 pixels. This error may seem important in absolute terms, but should nonetheless be put into perspective: it is expressed in pixels, but the FoE should not be treated in the same way as a simple interest point, since above all, it represents a measurement of a vehicle's different translation movements. In fact, if we consider the simulated optical system (which has a focal length of 10 mm and a pixel size of 10μm), an error of 1 pixel on the position of the FoE results in an error of 10^{-3} on the report: $\frac{\|T\|}{T_Z}$ where **T** is the translation vector. In the case of a vehicle traveling at 50 km/h, this represents a lateral translation of 0.6 mm. In our case, the error made on the location of the FoE therefore corresponds to an estimation error of 0.8% on the previous report. For comparative purposes, we also used a cumulative vote localization method, as shown in Suhr *et al.* (2008). This last method produces an average FoE localization error of 10.6 pixels, that is to say, an error of 1.6% on the estimate of the ratio $\frac{\|T\|}{T_Z}$.

Figure 2.11. *Image taken from a simulated sequence, the optical flow is calculated using the FOLKI method and the FoE using the reverse c-velocity*

Results on real images: the approach was tested on images taken from databases produced as part of the ANR LOVe project. The optical flow used was calculated using the FOLKI method (Le Besnerais *et al* 2005). It was not possible to use sensors accurate enough to provide an exploitable ground truth as a basis for comparison. At best, using data from the University of Karlsruhe, we can expect an estimate of the FoE's accurate position to be around 13 pixels, which is not enough to be able to form a solid basis for comparison. The evaluation of the quality of the FoE extraction by our approach was therefore done qualitatively on real images. The images in Figure 2.12 show the Focus of Expansion retrieved using our method for several real images. In any case, this FoE corresponds to the knowledge we have about the ego-vehicle's approximate movement. In addition, it may be interesting to consider the field lines of the optical flow as visual indicators of the quality of the extracted FoE.

Figure 2.12. *Example of results obtained on a real image*

Another qualitative indicator of the quality of the extracted FoE can be the comparison between the initial voting space (calculated with an assumed FoE at the center of the image) and the final voting space (calculated in relation to the extracted FoE). Such a comparison is illustrated in Figure 2.13. In both cases, it appears that after carrying out the optimization, the representation of the planes observed in the voting space is much more in conformity with the expected parabolic model, signaling that the estimated position of the FoE is closer to the real FoE after optimization, than before.

In conclusion, we have shown here how the use of the same technique of estimation of optical flow at the lowest level, made it possible to detect the 3D planes (and therefore the navigable space) step-by-step and obtain a rough estimate of the translational parameters of the ego-vehicle's 3D movement. The determination of the parallel frontal planes assimilated to obstacles, which will not be discussed here, could also benefit from the cumulative voting technique, but associating it with a space of parallel frontal planes. This would lead to proposing a monocular vision unified method, which can be adapted to the targeted objectives: road detection, obstacle detection or odometry. Limiting low-level perception operators could only contribute to the efficiency of embedded vision systems.

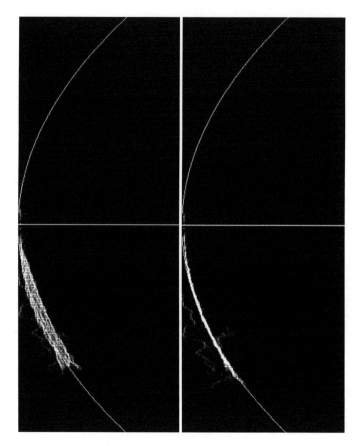

Figure 2.13. *Comparison of the voting spaces before (a) and after (b) FoE search. Dispersion is greatly reduced, thus enabling a more accurate estimation process of the 3D planes, since the latter offers a better estimation of the FoE (translational speed)*

2.3.5. *Detection of obstacles through the compensation of ego-movement*

This section presents a fine and accurate detection method for dynamic objects, using a moving stereovision sensor (Bak *et al.* 2010; Bak *et al.* 2011b). The approach remains fairly classic: estimating the proper movement of the ego-vehicle and then deducing – at the lowest level – the areas on the image showing an incompatible movement. Its salient feature is that it adds a fine estimation of the optical flow to this conventional phase, thus making it possible to extract the dynamic objects that are not consistent with the vehicle's movement, in a very accurate manner. The approach therefore combines a sparse matching of primitives and a more localized dense estimation, in order to obtain high accuracy in the detection. The extraction of the ego-movement is performed using two successive disparity maps and two images from one of the sensors. This amounts to considering not only a pair of stereo images, but also two successive images for one of the two cameras. A set of robust (SURF-type) interest points is extracted from each of the two images. The use of interest points makes it possible to envisage significant differences between two successive images. From this set of points and thanks to a RANSAC-type (RANdom SAmple Consensus) decision process, a system of equations resulting from the projection of the movement is solved. The use of the RANSAC enables certain insensitivity to the presence of false pairings, or to minor movements. The particularity of this technique is to do without the coordinates of the points in the 3D space. Indeed, while the noise on the image space can be entirely isotropic, this is not true for the object space. This method makes it possible to estimate the six movement parameters of the stereo sensor between two successive poses.

2.3.5.1. *Estimation of the ego-movement*

The suggested system is based on a pair of aligned and synchronized stereo cameras (where the optical axes are parallel). The road scene is represented by a set of two Cartesian reference frames. The first one, R_a, is absolute whereas the second one, R_r, is relative. At $t = t_0$, the two reference marks coincide. Their origin matches the main point on the right of the system.

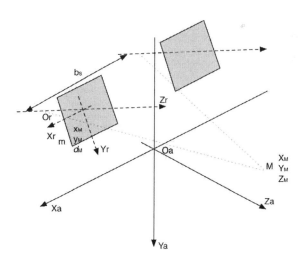

Figure 2.14. *Stereo system and cue used*

Let $M = \begin{vmatrix} X_M \\ Y_M \\ Z_M \end{vmatrix}$ be a 3D point in the road scene and $m = \begin{vmatrix} x_m = f\dfrac{X_M}{Z_M} \\ y_m = f\dfrac{Y_M}{Z_M} \\ \delta_m = f\dfrac{b_s}{Z_M} \end{vmatrix}$ its projection in

the image, with f being the focal length, b_s the base, and δ_m the disparity (right landmark). The disparity is estimated using classical method called *semi-global matching* (described in Hirschmuller (2005)), which is available in the OpenCV library.

Between t_0 and t_1, the binocular system has unconstrained movement $(\vec{T}, \vec{\Omega})$ where:

$\vec{T} = \begin{vmatrix} T_X \\ T_Y \\ T_Z \end{vmatrix}$ (resp. $\vec{\Omega} = \begin{vmatrix} \omega_X \\ \omega_Y \\ \omega_Z \end{vmatrix}$) is the translational (resp. rotational) component of 3D

motion. We assume that the components of $\vec{\Omega}$ are small enough to enable the linearization of the trigonometric lines. Under these conditions, the apparent motion in the image of a point m is:

$$\left\{ \begin{aligned} \mu &= \frac{xy}{f}\omega_X - \left(f + \frac{x^2}{f}\right)\omega_Y + y\omega_Z - \frac{\delta fT_X}{b_s} + \frac{x\delta T_Z}{b_s} \\ v &= \left(f + \frac{y^2}{f}\right)\omega_X - \frac{xy}{f}\omega_Y - x\omega_Z - \frac{\delta fT_Y}{b_s} + \frac{y\delta T_Z}{b_s} \\ \xi &= \delta\,\frac{y\omega_X - x\omega_Y + \dfrac{T_Z}{b_s}}{x\omega_Y - y\omega_X - \dfrac{T_Z}{b_s} + 1} \end{aligned} \right.$$

Its m' image is then: $m' = \begin{cases} x_m + \mu \\ y_m + v \\ \delta_m + \xi \end{cases} = P_{(\vec{T}, \vec{\Omega})}(m)$

2D equations of motion define a linear system in $(\vec{T}, \vec{\Omega})$. In order to solve this system, it is necessary to extract a S set of points mapped between the two stereo images. A matching process of SURF points then makes it possible to define an overdetermined system for which the estimation of $(\vec{T}, \vec{\Omega})$ amounts to minimizing an energy function: $\varepsilon = \sum\limits_{(m,m') \in S} dist\left(m', P_{(\vec{T},\vec{\Omega})}(m)\right)$, where *dist* is a convex metric. The estimation of the ego-movement then results in a least-squares optimization problem, which can be solved by using a classical approach, such as the Singular Value Decomposition (SVD).

Since the moving scene can contain both static objects (same movement as the estimated ego-movement) and dynamic objects, certain false matches are then considered as *outliers*, which may adversely affect the consistency of the estimates. This is the reason why a resolution based on the RANSAC approach is preferred.

This algorithm has been tested using the Pro-SiVIC simulator (ESI group and Gustave Eiffel University (COSYS-LIVIC)), among others. Over a sequence of 600 images representing an urban 250 meter-course, the error in final positioning made by the suggested system was inferior to 2 meters, for an instantaneous average error lower than 2%.

2.3.5.2. *Dynamic objects detection*

In order to detect dynamic objects in the scene, each point of the image is compared to the estimated movement model. Knowing the average error made by the extraction algorithm, we can estimate a neighborhood around each point, for which its correspondence at the next instant must be found. This point corresponds to a static object, thus making it possible to distinguish between static and dynamic objects. However, in order to be able to discriminate between two different motion objects, we choose to enlarge this neighborhood. Therefore, for each point of the image, a neighborhood for the following image is known. Its dimensions are defined by the accuracy of the ego-movement extraction algorithm, where we will find its correspondence. This search is done by using correlation calculations between the source point and all of the candidate points. In order to improve the method's robustness, a double selection is applied: among all of the candidates with a high score, the maximum scores are retained, while validating the reference point staticity hypothesis. This pairing stage makes it possible to construct a vector field, for which the various dynamic objects of the road scene will be able to be extracted. It is this vector field that is represented in false colors in the results images of Figure 2.15. Therefore, the system is based on a compensation for the ego-movement, and on the measurement of the 3D residual movement of the objects in the image. A segmentation of results makes it possible to provide high level information, while temporal integration ensures their temporal consistency. This temporal integration makes it possible to remove many false positives, while at the same time improving the detection of slow or distant objects.

Figure 2.15. *Example of detection of independent movements. For a color version of this figure, see www.iste.co.uk/bensrhair/adas.zip*

2.3.6. *Visual odometry*

2.3.6.1. *Omnidirectional image processing*

Due to the particular geometry induced by catadioptric systems, conventional image processing methods must be adapted to account for image distortion. Most of them use spherical representation to overcome the strong distortions of omnidirectional images (Geyer *et al.* 2002b; Bülow 2002). Several methods for detecting visual primitives in omnidirectional images have been proposed in literature. Hansen *et al.* (2007) proposed an adaptation of the SIFT detector/descriptor for unconventional images, and demonstrated that image processing adapted to the geometry of the sensor can significantly improve detection and matching performance. Hadj-Abdelkader *et al.* (2008b) and Demonceaux *et al.* (2011) have proposed works on the adaptation of a Harris detector for the geometry of unconventional cameras. Other methods for visual monitoring or optical flow estimation have been proposed, taking into account the geometry of the visual sensor (Rameau *et al.* 2011; Radgui *et al.* 2011).

2.3.6.2. Processing of spherical images in the spectral domain

In this section we will present a spherical image processing technique, using spherical harmonic decomposition. One of the central functions in image processing and the extraction of visual primitives is convolution-based filtering with a Gaussian filter. The first research studies dealing with omnidirectional images naively used filters designed for perspective images. However, the invariance of a classical filter in the planar image is no longer valid in a spherical image. Bülow (2004) proposed a spherical Gaussian filter as a solution to the representation of heat diffusion equations on a sphere. An example of the spherical Gaussian is shown in Figure 2.16.

Visual primitive extraction methods, such as point or edge detection are based on the derivative of the image. The latter is generally obtained thanks to a convolution with the Gaussian derivative. When the image is defined in the space of the sphere, the derivative of the spherical Gaussian must be used. Considering the representation of a spherical image $I(\theta, \varphi)$ through its latitude $\theta \in [0, \pi]$ and longitude $\varphi \in [0, 2\pi]$ angles, the derivative of the spherical image is given by:

$$\frac{\partial I(\theta, \varphi)}{\partial \theta} \approx \frac{\partial G}{\partial \theta} * I(\theta, \varphi)$$

$$\frac{\partial I(\theta, \varphi)}{\partial \varphi} \approx \frac{\partial G}{\partial \varphi} * I(\theta, \varphi)$$

where G is the spherical Gaussian defined at the north pole of the sphere. These derivatives are not easy to calculate because the spherical Gaussian G is invariant to longitude φ, and therefore $\frac{\partial G}{\partial \varphi} = 0$. Bülow (2002) proposes a solution to calculate the derivative of the spherical Gaussian with respect to longitude φ, described in three stages, as illustrated in Figure 2.17. The spherical Gaussian G is placed, initially, at the equator of the sphere via a rotation of $\pi/2$. Then, we calculate the Gaussian derivative in relation to the angle. Finally, the calculated derivative is replaced at the north pole of the sphere by the inverse of the previous rotation. The two derivative filters are applied by integration in SO(3) and despite this, the direction of the filter changes during the first rotation, before positioning the filter on the sphere by means of the two remaining rotations. This implies that the derivative of the spherical image is not performed correctly. Moreover, the derivative of the spherical image according to the angle θ is not considered in the work proposed in Bülow (2002).

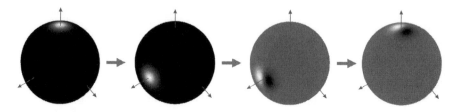

Figure 2.16. *Spherical Gaussian and its derivative, according to longitude φ. For a color version of this figure, see www.iste.co.uk/bensrhair/adas.zip*

2.3.6.3. *Derivative of a spherical image in the spectral space*

In order to calculate the derivative of a spherical image, depending on its latitude and longitude angles, the directional derivative of the spherical Gaussian is first calculated in relation to the angle φ, using the method proposed in Bülow (2002). Then, the derivative, depending on the latitude, is obtained by applying a rotation along the azimuthal axis of the sphere, from $\pi/2$. Let us note that all rotation operations are applied in spectral space, using spherical harmonics. The convolution is therefore performed by neglecting the first rotation along the azimuth of the sphere. Figure 2.17 illustrates the two Gaussian gradients depending on angles φ et θ and their convolutions. An example of the derivative of an omnidirectional image using the proposed approach, is shown in Figure 2.18. Let us note that this convolution can be achieved by integrating in the space of the sphere (S^2). However, the result of the convolution is defined in SO(3).

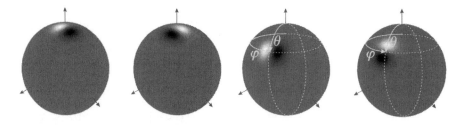

Figure 2.17. *Convolution with Gaussian gradients according to angles φ and θ. For a color version of this figure, see www.iste.co.uk/bensrhair/adas.zip*

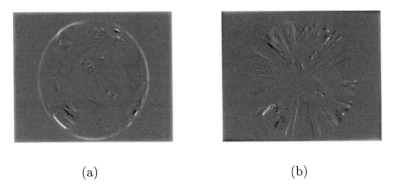

(a)   (b)

Figure 2.18. *Omnidirectional image derivatives: (a) by latitude,*
(b) by longitude. For a color version of this figure, see
www.iste.co.uk/bensrhair/adas.zip

2.3.6.4. *Experimental evaluation*

The calculation of the gradient of the spherical image is evaluated *via* the detection and matching of Harris points. This calculation is based on the image's gradient. The contribution of image processing in a space dedicated to the geometry of the camera, can be appreciated through a sparse visual odometry based on the calculation of the essential matrix between two shots. Visual odometry can be obtained in order to compare it with the ground truth, on one hand, and with the one obtained by means of image processing methods adapted to conventional cases (designed for perspective images), on the other hand. The experimental results illustrated below (Figures 2.19, 2.20 and 2.21) have been obtained for a sequence of omnidirectional images acquired by a catadioptric camera embedded into a mobile robot moving around in an indoor environment. The sequence is made up of 600 images of 1027x768 pixels resolution. The ground truth used in this evaluation is obtained by integrating odometric measurements drawn from the proprioceptive sensors of the mobile robot (sensors considered to be sufficiently accurate). The estimation of the pose of the camera with respect to the initial position (reference pose) is incremental and checked every 20 images. Let us observe that this *baseline* can be increased when a sufficient amount of Harris points has been matched. However, the pose calculation using conventional image processing is often unsuccessful. This is due to the strong distortions in the image following ample camera displacements.

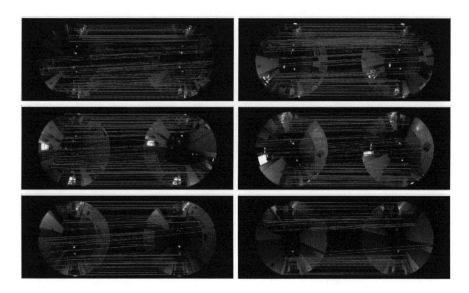

Figure 2.19. *Example of a matching of Harris points using the suggested method.*
For a color version of this figure, see www.iste.co.uk/bensrhair/adas.zip

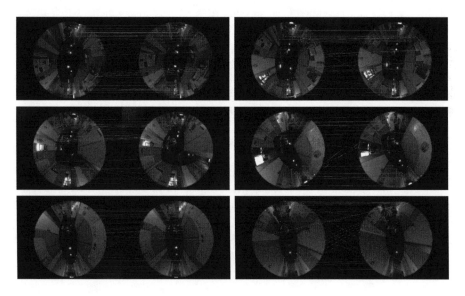

Figure 2.20. *Example of a matching of Harris points using the suggested method.*
For a color version of this figure, see www.iste.co.uk/bensrhair/adas.zip

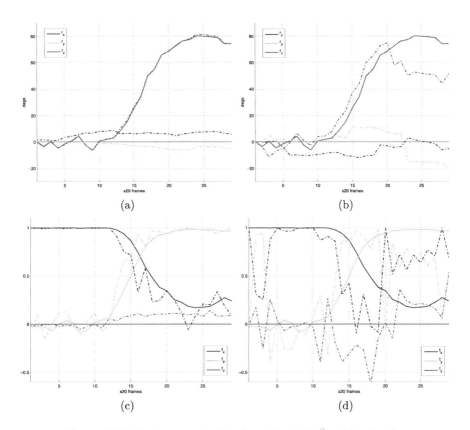

Figure 2.21. *Displacement estimation: (a) rotation and (c) direction of translation using the adapted approach, (b) rotation and (d) direction of translation using the classical approach*

2.4. Conclusion

Enabling an autonomous vehicle to detect the navigable space in which it evolves, at the same time as the potential obstacles, while being able to estimate its own ego-movement, is a complex task whose solutions provided so far are rich and can be combined, since they are often complementary. The biggest challenge is to propose sufficiently flexible and reconfigurable systems that are capable of using a limited number of vision operators, which are inexpensive in terms of computation time, while making the best use of the geometric and photometric characteristics of the visual sensors employed. This chapter has illustrated this approach by means of a choice of works relating to the detection of navigable space, the detection of moving objects and visual odometry. Without being exhaustive, this selection also shows that this type of approach makes it possible to make maximum the use of the information obtained from the visual sensors, and more particularly, from "low level" data. These methods also provide a sufficiently generic methodological framework to be exploited and used by new techniques, approaches and methods based on deep learning which, in turn, will attempt to break free from the use of models, if possible.

2.5. References

Aguilera, C.A., Aguilera, F.J., Sappa, A.D., Aguilera, C., Toledo, R. (2016). Learning cross-spectral similarity measures with deep convolutional neural networks. *Proceedings of the IEEE Conference on Computer Vision and Pattern Recognition Workshops*, 1–9.

Bak, A., Bouchafa, S., Aubert, D. (2010). Detection of independently moving objects through stereo vision and ego-motion extraction. *IEEE Intelligent Vehicles Symposium*, San Diego, CA, 863–870.

Bak, A., Bouchafa, S., Aubert, D. (2011a). Focus of expansion localization through inverse C-Velocity. *ICIAP*, (1), 484–493.

Bak, A., Bouchafa, S., Aubert, D. (2011b). Dynamic objects detection through visual odometry and stereo-vision: A study of inaccuracy and improvement sources. *Machine Vision and Applications*. Springer Verlag, Berlin.

Baker, S. and Nayar, S.K. (1999). A theory of single-viewpoint catadioptric image formation. *International Journal of Computer Vision*, 35(2), 175–196.

Barreto, J., Martin, F., Horaud, R. (2003). Visual servoing/tracking using central catadioptric images. In *Experimental Robotics VIII*, Siciliano, B. and and Dario, P. (eds). Springer Tracts in Advanced Robotics, Springer Verlag, Berlin.

Blanchon, M., Morel, O., Zhang, Y., Seulin, R., Crombez, N., Sidibé, D. (2019). Outdoor scenes pixel-wise semantic segmentation using polarimetry and fully convolutional network. *Proceedings of International Joint Conference on Computer Vision, Imaging and Computer Graphics Theory and Applications*, 328–335.

Bouchafa, S. and Zavidovique, B. (2012a). c-Velocity: A flow-cumulating uncalibrated approach for 3D plane detection. *International Journal of Computer Vision*, 97(2), 148–166.

Bouchafa, S. and Zavidovique, B. (2012b). Error sources and their impact on C-Velocity methods. *Pattern Recognition and Image Analysis*, 22(1), 168–179.

Braillon, C., Pradalier, C., Usher, K., Crowley, J., Laugier, C. (2008). Occupancy grids from stereo and optical flow data. *Experimental Robotics*, 39, 367–376.

Brox, T. and Malik, J. (2011). Large displacement optical flow: Descriptor matching in variational motion estimation. *IEEE Transactions on Pattern Analysis and Machine Intelligence*, 33(3), 500–513.

Brox, T., Bruhn, A., Papenberg, N., Weickert, J. (2004). High accuracy optical flow estimation based on a theory for warping. *Computer Vision – ECCV*. Springer, Berlin Heidelberg.

Bülow, T. (2002). Multiscale image processing on the sphere. *DAGM: Joint Pattern Recognition Symposium*. Springer, September 16–18, 609–617.

Bülow, T. (2004). Spherical diffusion for 3D surface smoothing. *IEEE Transactions on Pattern Analysis and Machine Intelligence*, 26(12), 1650–1654.

Chapoulie, A., Rives, P., Filliat, D. (2011). A spherical representation for efficient visual loop closing. *Proceedings of the IEEE International Conference on Computer Vision (ICCV) Workshops*, 335–342.

Chatfield, K., Lempitsky, V., Vedaldi, A., Zisserman, A. (2011). The devil is in the details: An evaluation of recent feature encoding methods. *Proceedings of the British Machine Vision Conference (BMVC)*, 76.1–76.2.

Courbon, J., Mezouar, Y., Eckt, L., Martinet, P. (2007). A generic fisheye camera model for robotic applications. *IEEE International Conference on Intelligent Robots and Systems*, 1683–1688.

Demonceaux, C., Vasseur, P., Fougerolle, Y.D. (2011). Central catadioptric image processing with geodesic metric. *Image and Vision Computing*, 29(12), 840–849.

Dornaika, F. and Chung, R. (2000). Cooperative stereo-motion: Matching and reconstruction. *Computer Vision and Image Understanding*, 79(3), 408–427.

Durrant-Whyte, H. and Bailey, T. (2006). Simultaneous localization and mapping: Part I. *IEEE Robotics and Automation Magazine*, 13(2), 99–110.

Eade, E. and Drummond, T. (2009). Edge landmarks in monocular SLAM. *Image and Vision Computing*, 588–596.

Engel, J., Schöps, T., Cremers, D. (2014). LSD-SLAM: Large-scale direct monocular SLAM. *Proceedings of European Conference on Computer Vision (ECCV)*, 834–849.

Engel, J., Koltun, V., Cremers, D. (2018), Direct sparse odometry. *IEEE Transaction on Pattern Analysis and Machine Intelligence*, 40(3), 611–625.

Fan, W., Ainouz, S., Mériaudeau, F. (2018). Polarization-based car detection. *Proceedings of IEEE International Conference on Image Processing*, 3069–3073.

Fermuller, C. and Aloimonos, Y. (1995). Qualitative egomotion. *International Journal of Computer Vision*, 15, 7–29.

Firmenichy, D., Brown, M., Susstrunk, S. (2011). Multispectral interest points for RGB-NIR image registration. *Proceedings of the 18th IEEE International Conference on Image Processing (ICIP)*, 181–184.

Fischler, M.A. and Bolles, R.C. (1981). Random sample consensus: A paradigm for model fitting with applications to image analysis and automated cartography. *Communication of ACM*, 24(6), 381–395.

Franke, U., Rabe, C., Badino, H., Gehrig, S. (2005). 6d vision: Fusion of stereo and motion for robust environment perception. *Lecture Notes in Computer Science*, 3663, 216–223.

Gallego, G., Tobi Delbruck, T., Orchard, G., Bartolozzi, C., Taba, B., Censi, A., Leutenegger, S., Davison, A., Conradt, J., Daniilidis, K., Scaramuzza, D. (2019). Event-based vision: A survey. *IEEE Transactions on Pattern Analysis and Machine Intelligence*, arXiv:1904.08405.

Geyer, C. and Daniilidis, K. (2000). A unifying theory for central panoramic systems and practical implications. *European Conference on Computer Vision*, 445–461.

Geyer, C. and Daniilidis, K. (2002a). Catadioptric projective geometry. *International Journal of Computer Vision*, 45, 223–243.

Geyer, C. and Daniilidis, K. (2002b). Properties of the catadioptric fundamental matrix. *Lecture Notes in Computer Science, Proceedings of the 7th European Conference on Computer Vision (ECCV)*, Part II, 140–154.

Geyer, C. and Daniilidis, K. (2003). Mirrors in motion: Epipolar geometry and motion estimation. *International Conference on Computer Vision*, 766–773.

Guo, J. and Brady, D. (2000). Fabrication of thin-film micropolarizer arrays for visible imaging polarimetry. *Applied Optics*, 39(10), 1486–1492.

Hadj-Abdelkader, H., Mezouar, Y., Andreff, A., Martinet, P. (2005). 2 1/2 D visual servoing with central catadioptric cameras. *International Conference on Intelligent Robots and Systems*, 3572–3577.

Hadj-Abdelkader, H., Mezouar, Y., Martinet, P., Chaumette, F. (2008a). Catadioptric visual servoing from 3-D straight lines. *IEEE Transactions on Robotics*, 24(3), 652–665.

Hadj-Abdelkader, H., Malis, E., Rives, P. (2008b). Spherical image processing for accurate visual odometry with omnidirectional cameras. *The 8th Workshop on Omnidirectional Vision, Camera Networks and Non-classical Cameras – OMNIVIS*, 25.

Hadj-Abdelkader, H., Mezouar, Y., Chateau, T. (2012). Generic realtime kernel-based tracking. *IEEE International Conference on Robotics and Automation*, 3069–3074.

Hansen, P., Corke, P., Boles, W., Daniilidis, K. (2007). Scale invariant feature matching with wide angle images. *International Conference on Intelligent Robots and Systems*, 1689–1694.

Harris, C. and Stephen, M. (1988). A combined corner and edge detector. *Proceedings of the 4th Alvey Vision Conference*, 147–151.

Heinrich, S. (2002). Fast obstacle detection using flow/depth constraint. *Proceedings of IEEE Intelligent Vehicle Symposium*, 658–665.

Hirschmuller, H. (2005). Accurate and efficient stereo-processing by semi-global matching and mutual information. *IEEE Computer Society Conference on Computer Vision and Pattern Recognition*, 2, 807–814.

Horn, B.K.P. and Schunck, B.G. (1981). Determining optical flow. *Artificial Intelligence*, 17(1–3), 185–203.

Labayrade, R., Aubert, D., Tarel, J.P. (2002), Real time obstacle detection in stereovision on non flat road geometry through "v-disparity" representation. *IEEE Intelligent Vehicle Symposium*, 2, 646–665.

Labayrade, R., Royere, C., Gruyer, D., Aubert, D. (2005). Cooperative fusion for multi-obstacles detection with the use of stereovision and laser scanner. *Autonomous Robots*, 19(2), 117–140 [Online]. Available at: DOI: 10.1007/s10514-005-0611-7.

Labayrade, R., Gruyer, D., Royere, C., Perrolaz, M. (2007). Obstacle detection in outdoor environments based on fusion between stereovision and laser scanner. *Mobile Robots: Perception & Navigation*. InTech, 91–110.

Lategahn, H., Beck, J., Kitt, B., Stiller, C. (2013). How to learn an illumination robust image feature for place recognition. *Proceedings of the IEEE Intelligent Vehicles Symposium (IV)*, 285–291.

Le Besnerais, G. and Champagnat, F. (2005). Dense optical flow by iterative local window registration. *IEEE International Conference on Image Processing*, 137–140.

Leibe, B., Cornelis, N., Cornelis, K., Van Gool, L. (2007). Dynamic 3d scene analysis from a moving vehicle. *IEEE Conference on Computer Vision and Pattern Recognition*, 1–8.

Lemaire, T. and Lacroix, S. (2007). Monocular-vision based SLAM using line segments. *Proceedings of ICRA*, 2791–2796.

Lenz, P., Ziegler, J., Geiger, A., Roser, M. (2011). Sparse scene flow segmentation for moving object detection in urban environments. *IEEE Intelligent Vehicles Symposium*, 926–932.

Longuet-Higgins, H.C. and Prazdny, K. (1980). The interpretation of a moving retinal image. *Proceedings of the Royal Society of London. Series B, Biological Sciences*, 208(1173), 385–397.

Lowe, D.G. (2004), Distinctive image features from scale-invariant keypoints. *International Journal of Computer Vision*, 60(2), 91–110.

Lucas, B.D. and Kanade, T. (1981). An iterative image registration technique with an application to stereo vision. *Proceedings of the 7th International Joint Conference on Artificial Intelligence*, 674–679.

Maddern, W. and Vidas, S. (2012). Towards robust night and day place recognition using visible and thermal imaging. *RSS 2012: Beyond Laser and Vision: Alternative Sensing Techniques for Robotic Perception*, University of Sydney.

Magnabosco, M. and Breckon, T.P. (2013). Cross-spectral visual Simultaneous Localization And Mapping (SLAM) with sensor handover, *Robotics and Autonomous Systems*, 61(2), 195–208.

Mai, T.K., Gouiffès, M., Bouchafa, S. (2017a). Exploiting optical flow field properties for 3D structure identification. *ICINCO*, (2), 459–464.

Mai, T.K., Gouiffès, M., Bouchafa, S. (2017b). Optical flow refinement using reliable flow propagation. *VISIGRAPP (6: VISAPP)*, 451–458.

Mai, T.K., Gouiffès, M., Bouchafa, S. (2020). Optical flow refinement using iterative propagation under colour, proximity and flow reliability constraints. *IET Image Process*, 14(8), 1509–1519.

Mei, C. and Rives, P. (2008). Single view point omnidirectional camera calibration from planar grids. *IEEE International Conference on Robotics and Automation*, 3945–3950.

Mei, C., Benhimane, S., Malis, E., Rives, P. (2008). Efficient homography-based tracking and 3D reconstruction for single-viewpoint sensors. *IEEE Transactions on Robotics*, 24(6), 1352–1364.

Mouats, T. and Aouf, N. (2013). Multimodal stereo correspondence based on phase congruency and edge histogram descriptor. *Proceedings of the 16th International Conference on Information Fusion (FUSION)*, 1981–1987.

Mur-Artal, R., Montiel, J.M.M., Tardos, J.D. (2015). RB-SLAM: A versatile and accurate monocular SLAM system. *IEEE Transaction on Robotics*, 31(5), 1147–1163.

Naseer, T., Spinello, L., Burgard, W., Stachniss, C. (2014). Robust visual robot localization across seasons using network flows. *Proceedings of the 28th AAAI Conference on Artificial Intelligence*, 2564–2570.

Nedevschi, S., Bota, S., Tomiuc, C. (2009). Stereo-based pedestrian detection for collision-avoidance applications. *IEEE Intelligent Transportation Systems*, 10(3), 380–391.

Neubert, P., Sunderhauf, N., Protzel, P. (2013). Appearance change prediction for long-term navigation across seasons. *Proceedings of the European Conference on Mobile Robots (ECMR)*, 198–203.

Newcombe, R.A., Lovegrove, S.J., Davison, A.J. (2011). DTAM: Dense tracking and mapping in real-time. *Proceedings of IEEE International Conference on Computer Vision (ICCV)*, 2320–2327.

Nocedal, J. and Wright, S.J. (2006). *Numerical Optimization*, 2nd edition. Springer, New York.

Nordin, G.P., Meier, J.T., Deguzman, P.C., Jones, M.W. (1999). Micropolarizer array for infrared imaging polarimetry. *Journal of the Optical Society of America A*, 16(5), 1168–1174.

Oliva, A. and Torralba, A. (2001). Modeling the shape of the scene: A holistic representation of the spatial envelope. *International Journal of Computer Vision*, 42(3), 145–175.

Plyer, A., Le Besnerais, G., Champagnat, F. (2016). Massively parallel Lucas Kanade optical flow for real-time video processing applications. *Journal of Real-Time Image Processing*, 11(4), 713–730.

Pons, J., Keriven, R., Faugeras, O. (2007). Multi-view stereo reconstruction and scene flow estimation with a global image-based matching score. *International Journal of Computer Vision*, 72(2), 179–193.

Radgui, A., Demonceaux, C., Moaddib, M., Rziza, M., Aboutajdine, D. (2011). Optical flow estimation from multichannel spherical image decomposition. *Computer Vision and Image Understanding*, 115(9), 1263–1272.

Rameau, F., Sidibé, D., Demonceaux, C., Fofi, D. (2011). Tracking moving objects with a catadioptric sensor using particle filters. *Proceedings of ICCV Workshops*, 328–334.

Rastgoo, M., Demonceaux, C., Seulin, R., Morel, O. (2018). Attitude estimation from polarimetric cameras. *Proceedings of IEEE International Conference on Intelligent Robots and Systems*, 8397–8403.

Ricaurte, P., Chilán, C., Aguilera-Carrasco, C.A., Vintimilla, B.X, Sappa, A.D. (2014). Feature point descriptors: Infrared and visible spectra. *Sensors*, 14(2), 3690–3701.

Rodriguez, F., Frémont, V., Bonnifait, P., Cherfaoui, V. (2010). Visual confirmation of mobile objects tracked by a multi-layer lidar. *IEEE International Conference on Intelligent Transportation Systems*, 849–854.

Scaramuzza, D. and Fraundorfer, F. (2011). Visual odometry (tutorial). *IEEE Robotics and Automation Magazine*, 18(4), 80–92.

Shi, J. and Tomasi, C. (1994). Good features to track. *Proceedings of IEEE Conference on Computer Vision and Pattern Recognition*, 593–600.

Soquet, N., Perrollaz, M., Aubert, D. (2007). Free space estimation for autonomous navigation. *5th International Conference on Computer Vision Systems*, 1–6.

Suhr, J., Jung, H., Bae, K., Kim, J. (2008). Outlier rejection for cameras on intelligent vehicles. *Pattern Recognition Letters*, 29, 828–840.

Sun, D., Roth, S., Black, M.J. (2010). Secrets of optical flow estimation and their principles. *Computer Vision and Pattern Recognition*, 2432–2439.

Sünderhauf, N., Neubert, P., Protzel, P. (2013). Are we there yet? Challenging SeqSLAM on a 3000 km journey across all four seasons. *Proceedings of the IEEE International Conference on Robotics and Automation (ICRA) Workshop on Long-Term Autonomy*, 2013.

Triggs, B., McLauchlan, P.F., Hartley, R., Fitzgibbon, A.W. (1999). Bundle adjustment: A modern synthesis. *International Workshop on Vision Algorithms*, 298–372.

Tuytelaars, T. and Mikolajczyk, K. (2008). Local invariant feature detectors: A survey. *Foundations and Trends in Computer Graphics and Vision*, 3(3), 177–280.

Verri, A. and Poggio, T. (1989). Motion field and optical flow: Qualitative properties. *IEEE Transactions on Pattern Analysis and Machine Intelligence*, 11(5), 490–498.

Vu, T., Burlet, J., Aycard, O. (2008). Grid-based localization and online mapping with moving objects detection and tracking: New results. *Information Fusion*, Elsevier, 12(1), 58–69.

Wang, B., Florez, S.A., Frémont, V. (2014). Multiple obstacle detection and tracking using stereo vision: Application and analysis. *13th International Conference on Control Automation Robotics & Vision (ICARCV)*, 1074–1079.

Wedel, A., Rabe, C., Vaudrey, T., Brox, T., Franke, U. (2008). Efficient dense scene flow from sparse or dense stereo data. *European Conference on Computer Vision*, 739–751.

Weyand, T., Kostrikov, I., Philbin, J. (2016). – plaNet – photo geolocation with convolutional neural networks. *European Conference on Computer Vision*, 37–55.

Automated Driving, a Question of Trajectory Planning

3.1. Definition of planning

Although widely used in robotics, the term "planning" is much older. Its use is widespread in economics, production and neuroscience. Interestingly, it is the latter that inspired the works presented in the rest of this chapter. Neuroscientist Adrian M. Owen (1997) defined planning as organizing one's behavior in space and time in a specific situation where a goal/objective is to be achieved; this is based on a definition close to the one by Jouandet (1979). Fulfilling one's schedule, organizing one's workspace or even arranging processes are three very different examples of planning applications. Subsequently, robotics added optimality issues to this definition, although humans already perform optimal planning in a certain way, following particular configurations. However, with the explosion of disciplinary and thematic denominations in science, strong overlaps between these topics have emerged. Thus, the optimal planning of trajectories of dynamic systems treated in a continuous way (continuous spaces, continuous dynamics, etc.) can, at any point, be similar to optimal control. In his book, *Planning Algorithms*, Steven M. LaValle (2006) does not make any distinction between planning and control, even though he concedes that many authors agree that there is hierarchical level difference between the two terms. In fact, planning is generally carried out before control. In the field of automotive and autonomous mobility, we consider that planning can be taken as a term used in a generic manner, partially encompassing control, but also trajectory planning and route planning (otherwise called path planning). Planning can therefore be classified into three hierarchical levels, corresponding to the three hierarchical levels defined by Michon, as in Figure 3.1 (Michon 1985):

Chapter written by Olivier ORFILA, Dominique GRUYER and Rémi SAINCT.
For a color version of all the figures in this chapter, see www.iste.co.uk/bensrhair/adas.zip.

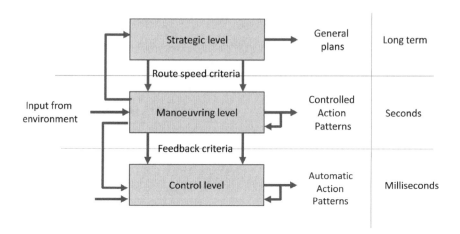

Figure 3.1. *Hierarchical structure of driving tasks (by Michon, based on the work of (Janssen 1979))*

– *Route planning*: corresponds to the strategic level. This level involves defining the itinerary choices and maneuvers at all the intersections on a route, starting from an origin and up to a destination. The order of magnitude of the distance between two required maneuvers is 200 to 1000 meters, although this parameter may be shorter in a city, or longer on the highway. This problem has been widely explored by the community, but there still remains a set of complex questions to be answered in relation to the specific problems of traveling salespeople, delivery agents or for the supervised management of vehicle fleets (such as taxis or professional fleets).

– *Trajectory planning*: corresponds to the tactical level. At this level, the challenge comes down to defining a path (determined at the "itinerary" planning level) which, among others, will contemplate the lane position, target speeds, acceleration, traffic lanes or even the gearbox ratio to be engaged. This level requires reacting to a higher frequency than mere route planning. The spatial range of action is estimated between 20 and 100 meters, or even less for events that require quicker reaction, as in the case of emergency collision avoidance.

– *Control planning*: corresponds to the operational level. The role of control planning is to determine the controls to be applied to the vehicle's actuator in order to follow the trajectory determined by the trajectory planning. This involves both reducing the error between the trajectory to be followed and the one actually completed, but also guaranteeing the stability of the dynamic system (vehicle).

3.2. Trajectory planning: general characteristics

As previously presented, trajectory planning constitutes an intermediate scale between route planning, going from an origin to a destination and control planning, which involves setting up the actuator controls which will make it possible to follow a given trajectory. Trajectory planning involves determining the path of the autonomous vehicle in space and in time, depending on a set of external constraints for the ego-vehicle (road infrastructure, other road users, obstacles, vulnerable users, traffic laws, weather conditions), or internal constraints (such as vehicle dynamics, actuator and perception capacities, the driver/passenger's state). In general, the simple problem of trajectory planning comes down to finding a *"feasible"* or *"applicable"* trajectory, that is to say, a trajectory that will respect a set of constraints and achieve its goal within a given time frame. For the purpose of our reasoning, we will assume that the solution space exists, and that the goal is physically attainable. However, most of the current works are devoted to optimal trajectory planning, with the intention of minimizing a cost function. Regarding these different types of planning, the obstacles currently observed and on which researchers are focusing their efforts are the following.

– The computation time in order to calculate multiple trajectories in real time (compatible with system dynamics and with the data processing that generates these trajectories). This poses a real challenge, as we know that at present, calculating a single trajectory in real time can already be difficult. This is particularly challenging for singular cases that require a rapid response from the system (reaction to a critical event), as in the event of planning a necessary emergency maneuver.

– System stability (implying high robustness), which has to be achieved and maintained regardless of the planned trajectory. Vehicle instability may result in a vehicle accident, and it is therefore necessary to ensure that this stability is guaranteed, taking into account the vehicle's dynamics and its tires.

– The asymptotic stability of a platoon of vehicles when several autonomous vehicles follow one another. In fact, in the case of a sudden change in the dynamics of the leading vehicle (leader), this situation could generate oscillations in speed (wave effect), which are increasingly important when the number of following vehicles increases.

– The optimality of the generated trajectories. Whether single or multi-objective, it is important for the generated trajectories not only to be feasible, but also optimal according to one or more goals, so that the autonomous vehicle can provide a solution to societal problems (pollution, safety, energy, economy, etc.). This optimality condition adds up a non-negligible computational complexity.

– The explainability of the planned trajectory (its ability to be explained), which can be complicated to obtain, for example, when Artificial Intelligence-based methods (neural networks, machine learning, deep learning, etc.) are involved. Explainability is important in order to be able to create a complete human-machine interface, and can also be used for solving liability conflicts in the event of an accident.

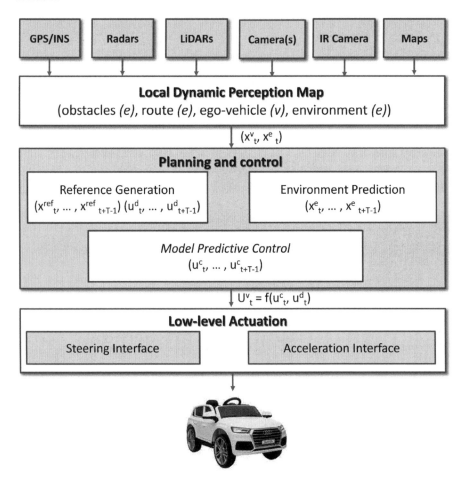

Figure 3.2. *Standard representation of the role of planning, here combining planning and control*

The first step for trajectory planning is to posit the problem. This problem has been posited in several manners, but the basic principles are always the same and involve defining: the system whose trajectory we wish to plan, optimization variables, the constraints applied to such variables, the goal(s) and their associated cost functions, and finally, the chosen planning method.

3.2.1. *Variables*

In this context, the important variables of the problem were judged as associated with the longitudinal and lateral behavior of the vehicle, but not including changes in gearbox ratio, even if the experimental vehicles in the laboratory are mechanical gearboxes. Indeed, vehicles using a mechanical gearbox are doomed to disappear due to the arrival of electric vehicles and the generalization of automatic boxes.

3.2.2. *Constraints*

The optimization problem constraints are those intrinsic to the problem, namely the limiting conditions associated with the system and, consequently, to the vehicle's dynamics, and the basic data of the problem (zero initial speed, etc.), as well as the limitations established by the Highway Code (traffic rules), which in this case were used as a reference for the safety assessment. If interaction with humans is contemplated, then constraints related to the driver's state can also be added.

3.2.3. *Cost functions*

After having defined the cost function resulting from the "objective" functions, it was necessary to simplify the driving model, summarizing the driver's behavior in order to limit the number of variables to be planned. For example, in the work of O. Orfila, the definition of eco-driving has been logically modeled using multi-objective optimization methods where the cost functions, for each objective, have also been modeled.

3.2.4. *Planning methodology*

There are many planning methods for indoor and outdoor mobile robotics. For the study of the latter – which focuses on the vehicles in a road environment – Benoît Vanholme (2012) and Laurène Claussmann (2019) both carried out a relatively complete literature review of existing methods for their research theses. A summary is provided in the rest of this chapter.

3.2.4.1. *Literature review*

The classification suggested for planning algorithms (Figure 3.3) is based on the following characteristics:

– the type of output: some algorithms provide one reference or trajectory solution (*"solve-algorithm"*), whereas others provide a set of solutions (*"set-algorithm"*). In the second case, a complementary algorithm is used to choose which trajectory should be followed;

– the algorithm's predictive (concerning decision-making and trajectory generation) or reactive aspect (modifying a trajectory in real time, using closed loop control);

– the mathematical field of the algorithm: geometric, heuristic, logical, cognitive or biomimetic.

The planning algorithms appearing in the literature can be classified into several large families. Figure 3.3 summarizes their main characteristics.

3.2.4.2. *Spatial configuration algorithms*

The algorithms from this family use the decomposition of the evolution space in a geometric way. They offer a set of solutions and are generally used for the generation of motions, or the deformation of a system. They can be used in a predictive manner, with a coarse mesh (space discretization) to limit the running time, or reactively, with a finer mesh.

Their main difficulties lie in obtaining the suitable spatial configuration parameters so as to correctly represent the movements and the environment: too coarse a discretization obscures (or blurs) collision risks disrespecting the kinematic constraints between two time steps; but too fine a discretization may hinder efficient use in real time.

In this family, we can distinguish:

– decompositions based on sampling, such as the *"Probabilistic RoadMap"* (PRM), where the trajectory is chosen from a random sample of points in the space, or sometimes in space-time;

– connected cell decompositions which can be exact, polar, use a dynamic window, or a *Voronoi* decomposition;

– lattice representations, which have the advantage of implicitly taking kinematic constraints into account, and can be adapted to the environment. However, these modelizations are more expensive in memory and in computation time.

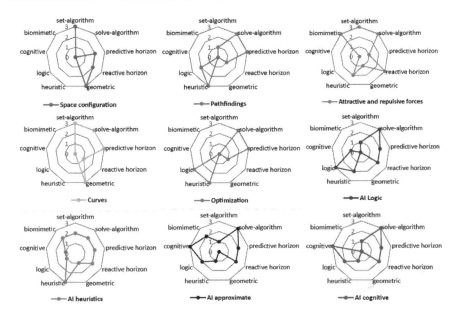

Figure 3.3. *"Radar" diagrams for the nine families of planning algorithms*

3.2.4.3. *The shortest path algorithms*

This family derives from the graph theory in operational search. The general idea is to find a path in a graph (a set of nodes and arcs, oriented or not), generated by a spatial configuration algorithm, while minimizing a cost function. The resolution is based on logical or heuristic methods, and can provide either a set of solutions or one reference solution.

These algorithms are most often used for route planning, but can also be used for local planning or state prediction. They are also suitable for unfamiliar environments. The best known are the *Dijkstra* and *A** algorithms, which will be presented in detail in section 1.3.1.1. Let us also quote *RRT* (*Rapidly-exploring Random Tree*) algorithms, similar to *PRM* (*Probabilistic Road Maps*) algorithms, but where the nodes are built from previous nodes, which guarantees kinematic feasibility while maintaining a reduced execution cost.

3.2.4.4. *Attractive and repulsive forces*

This biomimetic approach represents the space of evolution with attractive forces for the desired movements, such as speed, and repulsive forces, for obstacles. These algorithms have the advantage of being reactive to the dynamic evolution of the environment. The displacement of the ego-vehicle is then guided by the vector resulting from the different forces, without needing to discretize the solution space. They can provide a single solution or a set of movements; however, they are generally only used reactively, because the modeling of the whole evolution space has a high cost.

The Artificial Potential Fields (APF) method has been used in indoor mobile robotics and for very low speeds since 1978. Several extensions of this method have been put in practice in order to avoid oscillatory behavior and to get out of local minima, thanks to the addition of a heuristics. In 2017, Bounini proposed an adaptation of potential fields to manage the movement of a mobile robot in a closed workspace in real time. Virtual force fields (*VFF*) are also used to locally optimize a trajectory obtained by a coarser method, such as sampling for example. Application attempts to the road environment have been carried out, but cannot be generalized to all the configurations one may come across. For example, Wolf (2008) has adapted and used a set of four artificial potential fields for lanes, the road, obstacles and the desired speed. These potential fields modeled the functions to be applied to the highway. Sattel (2008) and Meinecke (2008) also propose approaches that can be considered as similar (*elastic bands*), to ensure lane keeping and anti-collision on motorways.

3.2.4.5. *Parametric and semi-parametric curves*

These curves are widely used for planning highway trajectories for two reasons: on the one hand, highways are a succession of simple and standardized geometric configurations (lines, circles and spirals), and on the other hand, this set of predefined configurations provide simple trajectories which can be quickly implemented and tested. The methods relying on them are generally used in a predictive manner for trajectory generation: they can either provide a reference solution, or a set of solutions.

Geometric considerations and dynamic constraints (speed profile, acceleration and braking capacity) are generally decoupled, and solved one after the other.

Within this family, we can distinguish:

– purely parametric models: lines, circles, spirals and sigmoid shapes, the latter two being respectively used for turns and lane changes;

– so-called semi-parametric modelings, used for joining a set of points (possibly predetermined by another method): polynomials, splines and B-splines.

3.2.4.6. *Numerical optimization algorithms*

Numerical optimization makes it possible to obtain a reference solution, by means of logical or heuristic methods. It is generally expressed as the minimization of a cost function for a set of state variables, within a set of constraints.

When this cost function and constraints (equality or inequality) are linear, we speak of linear programming, and minimization can be performed by the simplex algorithm, for example. We can also enumerate the problems of nonlinear regression, with the *Levenberg-Marquardt* algorithm, quadratic programming, predictive control algorithms, or dynamic programming.

These algorithms are widely used in trajectory planning, and are directly integrated in the form of a toolbox in various developments and programming software, such as Matlab for example, or in specially dedicated software.

3.2.4.7. *Artificial intelligence methods*

The main contribution of artificial intelligence to autonomous driving is its ability to reproduce and simulate the reasoning and learning abilities of drivers. For this reason, they are generally used for decision making.

AI techniques make use of very large databases, which can be incomplete and/or inaccurate. They can respond to generic questions and be modified without affecting the structure of the algorithm. AI brings together a wide variety of methods, which can be organized along two axes: rational or cognitive reasoning; and rule-based or learning-based action. We can therefore distinguish four sub-families: logical approaches, heuristic algorithms, approximate reasoning, and human-inspired methods.

– *Logic-based approaches* are expert systems (fact base, rule base, inference engine), which make it possible to solve specific complex tasks relying on a knowledge base. Their use is quick and intuitive, since each action has a well-defined cause, but their implementation requires an exhaustive list of rules and adjusting many parameters. Rules can be represented by decision trees or flowcharts. Finally, these systems can be heuristic, in particular for the management of uncertainties, for example, by using Bayesian networks or Markovian decision processes.

– *Heuristic algorithms* are used to quickly find an approximate solution when traditional exhaustive methods do not work. They often use an agent representation, as in the case of *Ant Colony Optimization*, presented in section 3.3.1.2. In general, evolutionary methods are inspired in the natural selection process, with stages of reproduction, mutation, recombination and agent selection. In this family we can also find all the learning methods, in particular support vector machines, for classification problems.

– So-called *approximate reasoning* AI methods emulate human reasoning. For example, fuzzy logic makes it possible to model approximate information (potentially linguistic variables modeling the knowledge of human experts on how a system works), and to combine several constraints with uncertainty. Expert systems relying on this logic are therefore more flexible, but lose their systematic methodology. Let us also mention artificial neural networks, which use the back propagation of the error to adjust the size of the connections between each neuron, using many variations in learning methods (supervised or not, by reinforcement, etc.).

– Finally, *human-inspired methods* bring together decision or evaluation processes which imitate human reasoning. In trajectory planning, the most widely used methods are risk estimators, for example, to assess the probability of a collision. There exist numerous estimators which can take into account the physiological state of the driver, acknowledging cognitive biases. This category also reunites interaction models based on game theory.

In the rest of this section, we will present a method developed at the COSYS department of the LIVIC laboratory, exploiting the findings made by Benoît Vanholme for his doctoral research.

3.2.5. *Co-pilot respecting legal traffic rules*

3.2.5.1. *Introduction*

As part of a European project (HAVEit) and a national project (ABV), Benoit Vanholme (2012) developed a highly automated driving delegation application, shared with a human driver. This co-piloting and optimal planning application (minimization of a set of criteria, including risk) tackled a multi-constraint problem.

First, it had to guarantee a high level of legal safety vis-à-vis the environment. That is to say, the virtual co-pilot had to abide by traffic rules and the Highway Code in order to guarantee safety and efficiency in mixed traffic conditions.

Second, its field of application had to be defined on an operating space guaranteeing user safety, and traffic rules being respected by every road user. In this area of operation, it was necessary to be able to implement all the necessary strategies in order to avoid collisions or, in the worst case, to mitigate them.

Finally, the human factor had to be included in the loop, and make it possible to share the driving task with the automaton. This active and/or informative human/machine interaction was clearly an unprecedented combination. It was therefore necessary to be able to efficiently manage a simple interface based on the *rider-horse* metaphor (H metaphor).

In terms of interactions between the driver and the co-pilot, the so-called "*cruise mode*" allowed for highly automated driving, taking into account the driver's wishes. In this way, the driver chooses the target speed as well as the target traffic lane, and the system fulfills this wish if it is legal, safe, and above all, physically feasible.

Then, in order to ensure user safety, a "*failsafe mode*" was proposed to stop the automated vehicle safely when – in the event of system failure – the driver is not able to effectively respond and/or react to the current driving situation.

Figure 3.4 presents the different modules, their interactions, the constraints applied to them (rule levels) and the actors involved in this highly automated and shared driving application.

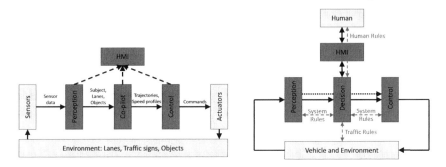

Figure 3.4. *Modules, rules, and interactions for the development of a co-pilot*

To produce the co-piloting module and to offer the control module at least one admissible trajectory, four steps are necessary. The first one involves retrieving the data resulting from perception, by estimating the attributes of the obstacles, the ego-vehicle, and the road (tracks and markings) within a common reference frame. Then, for the obstacles and for "*ghost*" obstacles, a first module generates the prediction of admissible and achievable trajectories. A second module generates the speed profiles and trajectories for the ego-vehicle. Finally, a last module applies traffic rules, human limitations and the system's constraints (perception/control) in order to evaluate and filter all the trajectories generated by the previous modules. In the end, at least one trajectory with a minimum cost will be retained. But before explaining the content of these three modules further, a first section will present all the rules applied as constraints to the speed profiles and to the predicted trajectories.

3.2.5.2. *The rules applied by the co-pilot*

The rule levels used are: the driving level (traffic rules: 1 to 9), the human level (rules 10 and 11) and the system's level (rules 12 to 15).

Rule 1: road users must avoid damaging road infrastructure or harming other road users.

Rule 2: the (human) driver must be in good physical and mental condition and must always be able to control the vehicle.

Rule 3: driving should be in the furthest right lane if possible, except for overtaking.

Rule 4: a vehicle must only be overtaken on the left lane, except in traffic jams, where overtaking on the right is permitted. An overtaking maneuver can only be engaged if the vehicles in front of and behind the ego-vehicle, on the same lane, have neither indicated nor started an overtaking maneuver of another vehicle. In addition, vehicles on the arrival lane must not be hampered by the maneuver of the ego-vehicle. An overtaking maneuver must not be carried out if it is prohibited by horizontal (continuous road markings) or vertical traffic signs. The corresponding flashing light must be activated all along the overtaking maneuver.

Rule 5: the speed must be adapted to road and weather conditions (for example, visibility and grip of the road and tires), speed limit signs and the presence of other vehicles. Inter-vehicle distance must be such that a collision can be avoided if a vehicle is performing emergency braking. The driver must also be able to avoid collisions with any "*predictable*" vehicle outside of their zone of perception.

Rule 6: braking should only be done for safety reasons and should be signaled by stop lights, which convey a braking maneuver.

Rule 7: only vehicles with a sufficiently powerful engine are allowed to travel on highways. Vehicles must not be driven in reverse or in the opposite direction. Vehicles already driving on the highway have priority over vehicles entering the highway. If the vehicle must be stopped for a technical reason, this should be done in the emergency stop lane, whenever possible.

Rule 8: the lighting and dynamics of the vehicle must be adapted to visibility conditions.

Rule 9: priority vehicles are exempt from traffic rules, except from rule 1.

Rule 10: in the "*Driver Only*" (DO) level of automation, the system is not active. In the "*Driver Assisted*" (DA) level of automation, the human driver performs longitudinal and lateral control, the driving system provides information on the optimum speed and the optimum lane. In the "*Semi Automated*" (SA) mode, the co-piloting system takes over longitudinal control. In the "*Highly Automated*" (HA) mode, the virtual co-pilot system takes longitudinal and lateral control of the vehicle, while the human driver monitors the situation and specifies the target speed and target lane. In the "*Fully Automated*" (FA) mode, the human driver no longer needs to monitor the vehicle's progress and maneuvers, and lane changes are automatically made. Optionally, each person can choose a driving style, for example "*normal*", "*sporting*" or "*comfortable*".

Rule 11: outside the area of application, only DO is possible. In the application area, the system changes from DO to DA. The automation mode can be changed by the human driver or the co-pilot system. The driver can switch between consecutive DA, SA, HA and FA automation levels. If the driver performs an action on the pedals or on the steering wheel, the automation level goes directly to DA. The system automatically switches from DA to SA in order to avoid a collision, by applying emergency braking. The system also switches to HA to avoid leaving the road. In the event of a failure in the co-piloting system or at the end of the application zone, the system automatically stops the vehicle on the emergency lane, unless the driver regains control of the vehicle.

Rule 12: in the perception zone, errors in the estimation of the attributes of the actors in the road scene (obstacles, road, ego-vehicle and environment) must be limited and must respect a minimum quality (Minimum Guaranteed Perception Quality).

Rule 13: the trajectories estimated by the obstacles and the ego-vehicle by the decision/planning module must be achievable by the controller. These trajectories must be physically valid.

Rule 14: the control module keeps the vehicle on a path with a bounded error. The accuracy of the control module is such that lane changes are made with a certain safety distance and, in extreme cases, the vehicle can stay on the target lane.

Rule 15: all the information communicated between components has a limited number of elements. Perception describes a maximum of three lanes: left, current and right. Perception describes a maximum of eight objects, as shown in Figure 3.5. These objects are the closest obstacles in front of and behind the ego-vehicle in each of the three lanes, and the objects on either side of the ego-vehicle. The decision module describes a maximum of four trajectories, an optimal trajectory for each lane and a "*safe*" trajectory which stops the vehicle if a failure occurs. In addition, the computation time of the perception, planning/decision and control modules must respect limited and predefined time periods. This last constraint must ensure real-time operation.

3.2.6. Trajectory prediction for "ghost" objects and vehicles

3.2.6.1. Calculation reference frame and track coordination system

In order to design this co-pilot, several steps were necessary. First of all, in order to simplify the stages for calculating trajectories and their level of risk, a change of reference frame was applied. We moved from the Cartesian XY "*world*" perception reference frame to a simpler, and above all, linear UW local reference frame. The UW curvilinear track coordinate system uses the same origin as the ego-vehicle XY coordinate system. In this new reference frame, the U axis is parallel to the middle of each lane and the W axis is perpendicular to U. This UW environment is a natural environment for trajectory calculations involving the ego-vehicle and the objects surrounding it. This UW lane coordinate system and the XY ego-vehicle coordinate system are shown in Figure 3.5. In the UW lane coordinate system, the centers of the traffic lanes have a constant W coordinate. The trajectories of the ego-vehicle and objects that target the center of the lane can be represented by two sections: a transient section (with variable W coordinates) and a permanent section (with constant W coordinates).

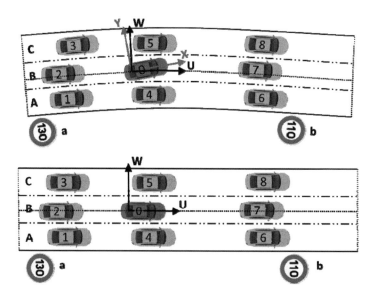

Figure 3.5. *Environment perception (ego-vehicle, obstacles, traffic lanes and markings), and passage from the real perception XY reference frame to the virtual planning UW reference frame*

Calculations with constant W coordinates are much easier and faster than calculations in real XY track geometry, which is usually (but not necessarily) based on a combination of lines, spirals and circles (Rajamani 2006). Subsequently, all ego-vehicle and object trajectory calculations will be made in UW.

3.2.6.2. Trajectory calculation for "ghost" objects and vehicles

The second stage will be to predict the legal safety trajectories on the three lanes (A, B and C) of the objects detected (objects numbered from 1 to 8), depending on their dynamic state and the traffic rules (Highway Code), applicable to the current configuration. These trajectories are visible in Figure 3.6.

To add to this concept of trajectory generation over a temporal space, an extension is proposed. This is presented in Figure 3.7 and illustrates the suggested concept off the "*mathematical area model*", generating the minimum and maximum trajectories (dotted line) which are achievable and admissible in relation to an exemplary real trajectory (full line).

In order to be able to guarantee maximum safety, even when distant perception is not available, the concept of a "ghost" vehicle has been proposed and developed. In fact, this new "safe" stage involves predicting safe trajectories which respect the traffic rules for the three traffic lanes (A–C) for the most "unfavorable" objects outside the perception zone (ghosts I to VI).

The next step will therefore predict safe speed profiles and respecting traffic rules for objects present in the environment close to the ego-vehicle (objects 1 to 8), and for ghost vehicles (I–VI) (Figures 3.8 and 3.9).

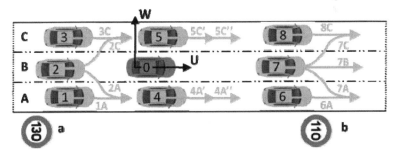

Figure 3.6. *Prediction of the possible trajectories of objects 1 to 8 depending on their position in relation to the ego-vehicle and traffic rules*

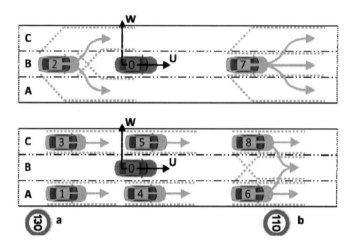

Figure 3.7. *Model of the evolution zones with predicted minimum and maximum trajectories (dotted line), compared to a real ideally predicted trajectory (full line)*

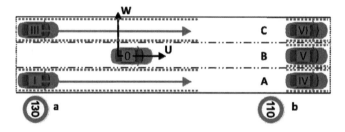

Figure 3.8. *Prediction of the trajectories of the ghosts (I–VI) depending on the position of the ego-vehicle, including the zones of minimum and maximum trajectories*

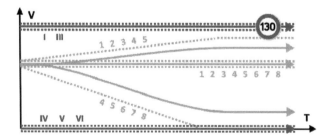

Figure 3.9. *Prediction of speed profiles for objects (1–8) and ghosts (I–VI) according to the ego-vehicle's position (0)*

3.2.6.3. *Prediction of the vehicle's speed profiles and trajectories*

Once the trajectories of real and ghost objects have been predicted, we must then calculate the safe speed profiles which respect to the above-mentioned rules for the ego-vehicle (0) and for the three traffic lanes (AC). These speed profiles for the ego-vehicle must meet the constraints of speed intelligent adaptation and the maintenance of safety inter-distances.

The calculation of the trajectories that can be reached by the ego-vehicle is strongly based on the use of an environment perception module (detection of obstacles and traffic lanes), and on the prediction of the trajectory of the objects already presented in the previous section. In the existing literature on trajectory planning, there are two main types of algorithms: "*sampling-based*" algorithms and direct algorithms (LaValle 2006). "*Sampling-based*" algorithms such as the "*sampling-based roadmap*", RRT algorithms or grid-based algorithms (space discretization) enable a universal approach, by first generating random samples in the space of trajectories, and evaluating these samples afterwards. Direct algorithms, such as those based on expert systems, potential fields or that are control-based, offer an application-specific approach which directly takes into account every driving aspect when generating a trajectory, without requiring an evaluation stage. Direct algorithms find more optimal solutions and require fewer calculations than "*sampling-based*" algorithms. "*Sampling-based*" algorithms solve complex problems which direct algorithms find difficult to solve.

In our case, the "*safe decision*" module (from the viewpoint of a driving rule) combines direct path planning and a "*sampling-based*" algorithm. The "*direct calculations*" part is used for longitudinal maneuvers and for calculating the ego-vehicle's speed profiles. Direct calculations are simple and accurate when using continuous variables, such as longitudinal speed from zero to maximum speed, or longitudinal acceleration, ranging from extreme braking to strong acceleration. Calculations using the "*sampling-based*" approach are used for lateral maneuvers, which are inherently unobtrusive. This is mainly due to the structure of the traffic lanes. In the current implementation of the decision module, only the trajectories centered towards the middle of the lanes are calculated.

Figure 3.10 shows the generation of seven possible speed profiles for the ego-vehicle, both longitudinal and lateral. These profiles are broken down into 3 categories. The first one corresponds to normal vehicle operation (0A, 0B, 0C). The second one takes into account singular situations such as breakdown and failure of the ego-vehicle (FA, FB, FC). The safety speed profiles generated in this case (F for *failure*), must enable safe stopping for all users. The last category corresponds to a situation involving a safe reaction to a dangerous event requiring safety braking, or a collision requiring emergency braking (JB). This can happen when an unequipped vehicle does not respect traffic and human rules. Figure 3.10 presents the model for the zone/minimum and maximum attainable speed envelope for trajectories 0A, 0B, FA, FB and FC.

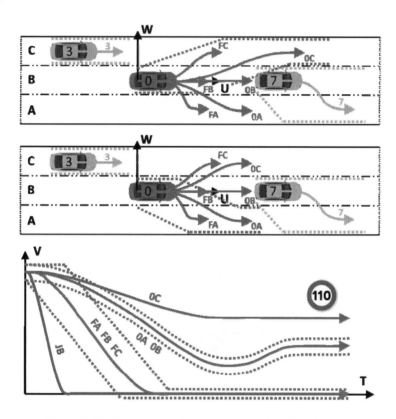

Figure 3.10. *Generation of the seven speed profiles (normal, failure, emergency braking) for the ego-vehicle*

The decision module first calculates the speed profiles before calculating the trajectories. This approach is opposed to the classic "*speed-path*" decomposition approach. In order to achieve the generation of the speed and acceleration profiles, we will implement a set of equations. These apply the constraints related to the limitations of the different parameters which have an impact on the ego-vehicle's dynamics (friction limits (G), human limits (H) and system constraints (I)) and depend on the infrastructure, the objects in the environment, and "*ghost*" vehicles, modeling borderline security cases imposed by the limitations of the electronic horizon produced by perception. These limitations are related to the vehicle's capacity, the driver's behavior and driving style and the limits of the perception and control modules. As a consequence, these limitations are highly dependent on the rules governing people as well as the system.

Equations [3.1] and [3.2] show the conditions for the transient section of a speed profile depending on longitudinal friction, as well as on human and system constraints. Equation [3.1] shows the most extreme deceleration.

$$\begin{cases} 0 \leq v_u \\ -min(a_u^G, a_u^H, a_u^I) = -a_u^J \leq a_u \end{cases} \qquad [3.1]$$

Equation [3.2] calculates a speed profile with maximum acceleration, making it possible to reach maximum speed. Maximum speed v_u is obtained from the target speed set by the human driver and the maximum speed for which the system is designed. For speed profiles in the "*breakdown*" and "*failure*" modes (FA, FB and FC), the target speed recommended by the driver is forced to 0 m/s and the acceleration produced by the driver is replaced by a maximum deceleration value which can be chosen between $-a_u^J$ and 0 m/s². a_u^G is the maximum acceleration for the available grip, a_u^H is the maximum acceleration as supported by humans and a_u^I is the maximum acceleration of the system. For the JB speed profile in "*emergency braking*", the driver's acceleration is replaced by an extreme deceleration value $-a_u^J$. In fact, for the JB profile, equations [3.1] and [3.2] are the same.

$$\begin{cases} v_u \leq min(v_u^H, v_u^I) \\ a_u \leq a_u^K = min(a_u^G, a_u^H, a_u^I) \end{cases} \qquad [3.2]$$

Equation [3.3] presents the conditions of the speed profile as a function of the lateral grip in the case of a curved track generating a centrifugal lateral acceleration. This equation also takes into account human limitations as well as the system's constraints. The lateral acceleration is a function of the longitudinal acceleration via the friction ellipse which models the level of grip between the road surface and the vehicle's tires. In equation [3.3], we can observe that the target speed is a function of the maximum lateral acceleration a_w^K and road curvature ρ^L. On a straight road, curvature tends towards 0 and the target speed tends towards an infinite value. When approaching a turn (curvature > 0), the speed of the ego-vehicle must be reduced to converge towards the recommended speed v_u. This action must be performed before reaching the start of the curve, at a distance P_u^L from the ego-vehicle. In this equation, t^R represents the reaction time of the system during this action.

$$\begin{cases} v_u \leq v_u^K = \sqrt{\dfrac{a_w^K}{\rho^L}} \\ a_u \leq -\dfrac{1}{2} \cdot \dfrac{(v_{u0})^2 - (v_u^K)^2}{P_u^L - v_{u0} \cdot t^R} \end{cases} \qquad [3.3]$$

The conditions applied to the speed profiles respecting the traffic rules are given by the following three equations. Equation [3.4] makes it possible to adapt the speed of the ego-vehicle to traffic rule 5. Equation [3.4] is equivalent to [3.3]. Traffic rule 5 states that the ego-vehicle must have a speed that will enable it to avoid a collision with a "*ghost*" vehicle.

$$\begin{cases} v_u \le v_u^S \\ a_u \le -\frac{1}{2} \cdot \frac{(v_{u0})^2 - (v_u^S)^2}{P_u^S - v_{u0} \cdot t^R} \end{cases}$$ [3.4]

That is, the target speed of the speed profile must allow braking with deceleration $-a_u^P$ for an object potentially present at position P_u^P, corresponding to the end of the perception zone. From this distance, we usually add an additional minimum safety distance d_u^J. Deceleration $-a_u^P$ must then be considered between $-a_u^J$ and 0. Higher values of a_u^P allow for higher ego-vehicle speeds, but cause more brutal braking when braking near an obstacle.

Solving the equation $v_u \cdot t^R + \frac{(v_u)^2}{2 \cdot a_u^P} = P_u^P - d_u^J$ at a speed v_u results in the fifth equation. This equation is identical for the 7 speed profiles of the ego-vehicle. In this model, "*ghost*" vehicles only represent "*potentially present*" obstacles. This concept does not represent the worst case of infrastructure configuration (curves and speed limits). However, the ego-vehicle's speed must still be limited in order to allow for a physically achievable and acceptable deceleration process within a constrained time-set, in case of "*ghost*" curves and "*ghost*" speed limits. This remark reveals an additional constraint: we must consider the hypothesis that the horizon of perception P_u^P and deceleration $-a_u^P$ for "*ghost*" curves and "*ghost*" speed limits are the same as for "*ghost*" obstacles. In this case, "*ghost*" vehicles at zero speed constitute a priority constraint to be taken into account when generating speed profiles.

$$\begin{cases} v_u \le a_u^P \cdot \left(-t^R + \sqrt{(t^R)^2 + 2 \cdot \frac{P_u^P - d_u^J}{a_u^P}} \right) \\ a_u \le a_u^K \end{cases}$$ [3.5]

Equation [3.6] describes the constraints applied to the speed profiles as a function of the existence of obstacles in front of the ego-vehicle. In this equation, an object (O) denotes any potential obstacle with a trajectory towards the same lane center as the ego-vehicle. In fluid traffic conditions, this equation also includes obstacles with trajectories towards an adjacent (left) lane to the speed profile's target lane, in order to avoid overtaking on the right. Obstacles on an adjacent traffic lane and at the rear of the ego-vehicle are not considered during the trajectory generation stage, but will be taken into account during the trajectory relevance evaluation stage.

$$\begin{cases} P_u^H = d_u^H + v_{u0}.t^H \\ P_u^I = d_u^I + v_{u0}.t^I + v_{u0}.t^R + \frac{(v_{u0})^2}{2.a_u^J} - \frac{(v_{u0}^O)^2}{2.\mu^L g} \\ P_u^K = max(P_u^H, P_u^I) \\ v_u \le v_u^O \\ a_u \le k_p.(P_{u0}^O - P_u^K) + k_v.(v_{u0}^O - v_{u0}) + a_{u0}^O \end{cases} \qquad [3.6]$$

In this equation, the *"target"* P_u^K distance to obstacles takes into account the P_u^H distance desired by the human driver and the safety distance P_u^I required by the system's constraints (perception and control). P_u^H is generally proportional to the ego-vehicle's speed. Time advance t^H is very much related to the driving style choice. For example, choosing a sporting driving style will produce shorter inter-vehicle distances (ego-vehicle and obstacles). This will then lead to a more abrupt and nervous driving and, by extension, to harder decelerations in the event the obstacle brakes. Distance P_u^I depends on the system's constraints. Rules 4 and 5 stipulate that this safety distance must be adapted to trigger collision avoidance in the event of emergency braking of a frontal obstacle. Therefore, equation [3.6] enables the system to keep a safe distance P_u^I, so that a small distance $d_u^I + v_{u0}.t^I$ is left when the obstacle brakes with a maximum deceleration of $-\mu^L.g$ (with $-\mu^L$, road grip) and the ego-vehicle braking with a maximum deceleration of $-a_u^J$, after reaction time t^R. The minimum system safety distance P_u^I mainly corresponds to the reaction time of the system t^R, when we make an abstraction from the first two terms (the safety margin when the ego-vehicle and the obstacle stop after emergency braking) and the fourth term (which only applies if the ego-vehicle and the obstacle have different braking capacities). The system's reaction time t^R is generally smaller than 1s, which is comparable to the reaction time of a human driver. This means that the system does not maintain greater inter-vehicle distances than those applied by a human driver.

Equations [3.1] to [3.6] presented constraints on the ego-vehicle speed profiles. The following equations will now introduce the constraints on the ego-vehicle trajectories. As for the speed profiles, the trajectories of the ego-vehicle must respect road grip limits, limits related to the human driver, and constraints imposed by the embedded system (perception and control). In this new set of equations, the first equation presents the slope for the *"comfortable"* trajectory mode S_w^K. If the presence of an obstacle in front of the ego-vehicle (equation [3.9]) does not impose any additional constraints, then the slope of the ego-vehicle trajectory is equal to S_w^K.

$$max(S_w^H, S_w^I) = S_w^K \le S_w \qquad [3.7]$$

The following equation describes the maximum slope for trajectory S_w^J. This value integrates the limits of the human S_w^H and of the system S_w^I.

$$S_w \leq S_w^J = min(S_w^G, S_w^H, S_w^I) \qquad [3.8]$$

Equations [3.1] to [3.8] represent constraints both on the speed profiles and on ego-vehicle trajectories. In the case of an overtaking maneuver, it will be necessary to adapt both the speed profile and the trajectory. As with a target following maneuver, changing lanes also involves the use of a safety distance P_u^J. This safety distance makes it possible to avoid an accident if an obstacle (on the current lane or the target lane) performs emergency braking. The last equation simply indicates that the slope of the trajectory S_w, the target speed profile v_u and acceleration a_u must meet certain constraints S_w^L, v_u^L and a_u^L so that the safety distance P_u^J is observed throughout the maneuver. S_w^L, v_u^L and a_u^L cannot be directly solved analytically. It has to be solved numerically, by sampling S_w^L, v_u^L and a_u^L, and checking the condition on P_u^J during the maneuver.

$$\begin{cases} v_u \leq v_u^L \\ a_u \leq a_u^L \\ S_w^L \leq S_w \end{cases} \qquad [3.9]$$

3.2.7. *Trajectory evaluation*

As we have seen previously, the decision module directly integrates most of the legal and regulatory aspects of safety when generating speed profiles and ego-vehicle trajectories. The speed profiles meeting the safety constraints are obtained by taking into account the minimum of speed profiles in relation to the constraints of equations [3.1] to [3.6]. The legal safety trajectories are found by considering the maximum (absolute values) of trajectories in relation to the constraints of equations [3.7] to [3.9]. This leads to seven speed profiles and seven trajectories for the ego-vehicle: 0A, 0B and 0C for normal system operation, FA, FB and FC for a "*breakdown*" and "*failure*" operation, and JB for collision mitigation and emergency braking. The goal of the evaluation stage will be to evaluate these seven trajectories on aspects that were not taken into account during the previous trajectory generation stage.

A performance cost is assigned to each of the remaining security aspects. The performance cost is binary, except for collisions with obstacles. In this case, the cost is proportional to the impact speed of the collision. This makes it possible to choose the trajectory with minimal collision impact in situations where an accident cannot be avoided. In the trajectory generation step, only "*ghost*" vehicles and frontal objects were taken into account. The trajectory evaluation stage also considers the prediction of the trajectories of ghosts and vehicles present at the rear of the ego-vehicle, as well as on the adjacent lanes. This evaluation stage verifies that the speed and speed profile of the ego-vehicle let "*ghost*" vehicles brake until they reach the ego-vehicle's speed with a reasonable deceleration. The evaluation stage also checks that the vehicles at the rear and on the sides do not need to brake during an ego-vehicle maneuver (rule 4). The evaluation only applies to lane changing trajectories 0A, 0C, FA and FC, and not to lane-keeping trajectories 0B, FB and JB. On the ego-vehicle's lane, this has priority over ghosts and vehicles behind it and on adjacent lanes.

The evaluation stage also takes into account the type of target lane and the associated markings to validate 0A, 0C, FA and FC lane changing trajectories (or not). Rule 7 states that, in normal operations, 0A and 0C trajectories should only target accessible lanes which do not have continuous lane markings. However, "*failure*" trajectories (FA and FC) can operate on an emergency stop lane or on a normal lane, and the ego-vehicle can then go through continuous lane markings. Indeed, changing lanes to an emergency stop lane is preferable to a collision with another vehicle. This maneuver is a priority safety maneuver.

Applying each rule greatly reduces the trajectory solution spaces. For example, traffic rules may exclude the possibility of changing lanes, human rules establish a limit for the ego-vehicle's speed, and system rules restrain the ego-vehicle's deceleration and acceleration capabilities. Nevertheless, after this evaluation, a minimum of at least one trajectory must exist in order to permit the safe evolution of the ego-vehicle.

3.2.8. *Results on real vehicles and on simulators*

In order to test, evaluate and validate the performance and capacities of this co-pilot, a set of use cases have been implemented in both real and virtual environments. The purpose is to confirm that the co-pilot is able to guarantee reliable, robust and, above all, safe operating and driving behavior in relation to different rule categories. The different scenarios used and tested were "*approaching a speed limit*", "*approaching a vehicle or a ghost zone*", "*following a target*", and "*overtaking a vehicle*".

It is important to note that this virtual co-pilot was prototyped and developed using the pro-SiVIC platform (Chapter 4), coupled with the Intempora RTMaps platform. In order to be able to concentrate solely on the development of the virtual co-pilot, the perception section was generated by the pro-SiVIC "*observers*". These "*observers*" are "*ground truths*" sensors which provide data and by extension, perfect perception. In order to validate the reliability and robustness of the suggested method, we have run scenarios with five and ten vehicles, each with its own co-pilot configured for one of the three driving modes ("*comfortable*", "*normal*", "*sporting*"). On the HMI (Human Machine Interface) presented in Figure 3.11, we observe the use of an instruction matrix for possible maneuvers and reachable zones (size 3*3). The first cell represents a left lane change and acceleration. The second cell represents a simple acceleration. The fifth cell represents a constant state (constant position and speed). The ninth cell represents a right lane change and deceleration. As we can see, the cells in red represent impossible maneuvers (trajectories, speed profiles) after the evaluation stage.

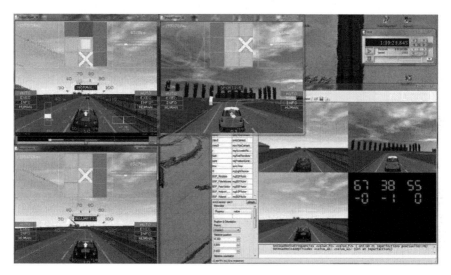

Figure 3.11. *Multi-vehicle and multimode piloting scenario (comfortable, normal and sporting)*

Once the results obtained in simulation were considered to be of sufficiently good quality, the co-pilot application was loaded into one of the LIVIC prototype vehicles dedicated to automated driving systems (CARRLA prototype) (Vanholme 2011). For this real conditioning, perception was obtained with the use of the modules and functions presented in Gruyer (2013) and Revilloud (2013).

Figures 3.12 to 3.14 present the results obtained for the first three use cases of embedded real prototypes, used on the tracks at Versailles-Satory. In Figure 3.12, we can clearly see the adaptation of the vehicle speed to a constraint (new speed limit). Figure 3.13 shows the reaction of the co-pilot recognizing a "ghost" vehicle after having detected an obstacle. Figure 3.14 shows a target tracking scenario, with the target making gear changes and one lane change.

Figure 3.15 shows the last use case in simulation, involving both longitudinal and lateral maneuvers. Figure 3.11 shows the application of the co-pilot at a complex scenario including five vehicles. The three visible vehicles are each following a different driving mode.

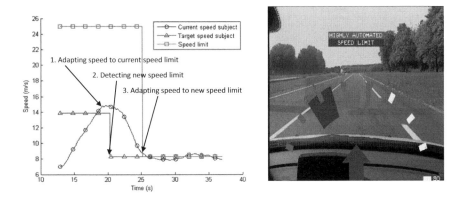

Figure 3.12. *"Approaching a speed limit" scenario*

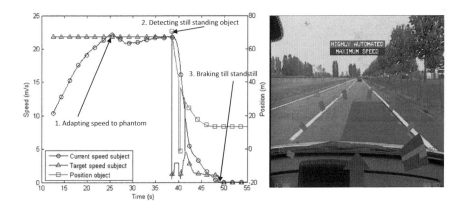

Figure 3.13. *"Approaching a vehicle or a ghost zone" scenario*

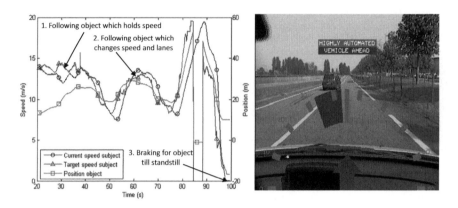

Figure 3.14. *"Following target" scenario, with a change in speed and lane*

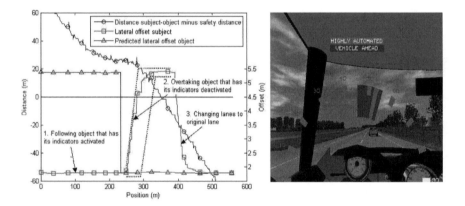

Figure 3.15. *"Overtaking a target and returning to the original lane" scenario*

3.3. Multi-objective trajectory planning

As presented previously, the goal of trajectory planning is to determine a feasible or ideal trajectory within a decision space (also called a solution space) determined by the perception (obstacles, traffic lanes and road markings, level of visibility), location and communication technologies of the autonomous vehicle. This decision space is a multidimensional space in which each axis is a variable in the planning problem.

Multi-objective trajectory planning, on the other hand, tries to find a compromise when several conflicting objectives must be managed at the same time. This is exactly the case faced by any human driver who must simultaneously deal with their own safety (which is also a constraint), their comfort and that of the passengers, their fuel consumption and travel time. Of course, other considerations – such as noise – can be taken into account, noise being perceived as a (dis)comfort indicator, but these will be deemed secondary to the previous ones.

As these goals are simultaneously managed by each driver, a hypothesis could be that each driver somehow allocates different weights to each of these goals and prioritizes them. Another hypothesis could be that drivers cannot handle more than one goal at a time, but this seems less likely, so we chose to focus on the first hypothesis.

Figure 3.16 illustrates how the weights allocated to the different goals (comfort, energy saving, safety and travel time) by human drivers can be distributed according to different driving styles (Orfila 2011). Actually, this distribution of optimization weights is even more dynamic than the figure suggests, knowing that an eco-driver can adopt a sporting driving profile if they believe they are in danger, or if they, as an exception, have to drive faster (planned delay at destination). These weights vary continuously during a journey – depending on the context – and the driving style can therefore be evaluated as a function of the average weight distribution over all the journeys. In Figure 3.16, we see that a "*hypermiler*" type of driver simply wishing to save fuel, focuses all their attention on this goal, abandoning safety or comfort goals. Conversely, a sporting driver will focus on travel time.

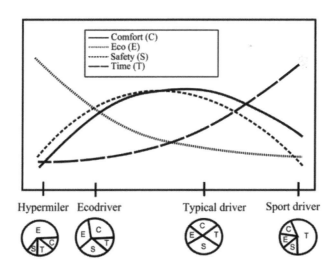

Figure 3.16. *Theoretical optimization weight distribution curves during multi-objective trajectory planning by a human driver*

The hypothesis assumed in these works is that planning multi-objective trajectories is a multi-objective optimization process, similar to the one performed by humans when driving. Indeed, multi-objective optimization carried out in real time might make it possible to reproduce the behavior of humans in the way they manage to cope with a multiplicity of goals while driving. Multi-objective trajectory planning can be achieved in many ways. The basic principle is – after having correctly posited the problem by choosing the variables to be controlled, their constraints and the cost functions to be optimized – to use a method that combines different optimization goals. Each of these goals will therefore be associated with a cost function. In this context, a new space is used for evaluating the solutions; the goals space. Like the decision space, this goals space is multidimensional, and each of its dimensions represents one goal in the planning problem. Figure 3.17 shows how each solution, represented by its variables in the decision space, has an equivalent in the goals space. The functions making it possible to transform a solution in the decision space into a solution evaluated in the goals space are precisely the cost functions of the multi-objective optimization problem.

Solving the multi-objective planning problem amounts to determining the set of optimal solutions for the optimization problem. Indeed, in a multi-objective problem, there is not one, but an infinity of optimal solutions which are all placed on the *Pareto* frontier (within the efficiency framework, as defined by *Pareto*).

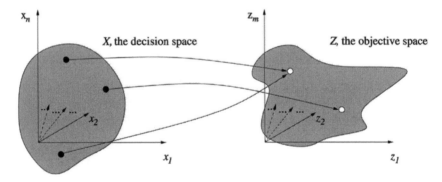

Figure 3.17. *Spaces used for multi-objective trajectory planning: on the left, the decision space and on the right, the objective space*

The *Pareto* frontier is the frontier over which the set of non-dominated solutions for the optimization problem is located. A solution is considered not dominated when at least one of its goals is inferior to that of the other solutions. Finding this *Pareto* frontier can be done in many ways. In his work on multi-objective optimization, Kalyanmoy Deb (2009) defines two main classes of approaches to solving multi-objective problems:

– Scalarization approaches, where we combine all the cost functions in one, in order to reduce them to a single-objective problem. This presupposes having varying parameters and requires one to repeat the optimization process in order to find all the *Pareto*-optimal solutions. Many scalarization variations exist, the main ones being:

- linear scalarization, which combines cost functions linearly in order to reduce them to a global cost function,

- nonlinear scalarization, which combines the cost functions in a nonlinear way in order to reduce them to a global cost function. This requires having a certain expertise of the issues dealt with, in order to be able to define these functions,

- the Epsilon constrained method, which constrains all the cost functions at a precise point, except a cost function we will try to minimize. Thus, we have a unique cost function, but we will have to repeat the optimization operation a certain number of times, by varying the values of the cost functions that we fix, in order to build the entire *Pareto* frontier.

– In so-called "*ideal*" approaches, an optimization method is set up for generating several solutions at the same time, until one converges towards the *Pareto* frontier. Here, the quality of each solution is generally estimated based on two main criteria: dominance and the density of the surrounding solutions.

3.3.1. *Linear scalarization*

The principle of linear scalarization is that a single and global cost function is determined as a linear combination of the cost functions depending on each of the goals. For a cost function f, we thus obtain the following equation:

$$f = \sum_{i=1}^{n} \alpha_i . f_i \qquad\qquad [3.10]$$

This assumes that, in order to find the set of optimal solutions, the optimization process must be repeated by varying α coefficients. Also, the complexity of linear scalarization is $O(2^n)$. Moreover, due to its linear description, this method does not allow for the construction of a *Pareto* frontier when the latter is not convex. However, its ease of use and interpretation make it a preferred solution which has been approached using several optimization techniques, including operational search (*A**, *Dijkstra*), ant colony optimization (ACO) and parametric methods. In the remainder of the chapter, we will introduce these three families of optimization algorithms applied to linear scalarization.

3.3.1.1. *Operational search*

Operational search is the whole of rational methods used for providing decision-making support. Among these methods, a subset of methods performs a graph search. In our case, we must first perform a decomposition of the searching space in order to obtain a graph, and then perform a search within this graph. These methods are often used in combinatorial optimization on problems such as the traveling salesperson problem.

In our case, we used this kind of method by breaking down the space – defined by a representation of the speed as a function of the distance – into square cells, keeping one distance step and a constant speed. These two parameters are essential parameters of the algorithm which easily make it possible to reduce the computation time but which, on the other hand, may negatively impact the quality of the solution. Although the decomposition into square cells has an undeniable advantage over the algorithm's simplicity (and its efficiency in terms of computing time), it presents a major drawback for the accuracy of results. In fact, in a description where we have the speed of the vehicle on the ordinate axis and the distance traveled on the abscissa axis, the same cell at low speed will represent a larger time than the same cell at high speed.

Among the different search algorithms in a graph, several solutions have been tested in the literature and two of the first to have been developed will be presented here: the *Dijkstra* algorithm, developed in 1959 (Dijkstra 1959) and the *A** algorithm, designed in 1968 (Hart 1968). In order to understand these two algorithms it is necessary to introduce the operating principle of the *Dijkstra* algorithm:

Step 1: start by initializing a node $w(A) = 0$ as the node's initial weight and $w(x) = \infty$ for all the other nodes, where x represents all other nodes.

Step 2: find the node x connected to the original node which has the weakest $w(x)$ weight. Stop the algorithm if $w(x) = \infty$ or if there are no more nodes. Node x then becomes the current node.

Step 3: for each node adjacent to x and identified as y, we must calculate:

if $w(x) + W(xy) < w(y)$, then $w(y)$ is updated as $w(x) + W(xy)$, where W is the cost for moving to the adjacent node. Then add x as a parent of y.

Step 4: repeat from step 2, until the shortest path is obtained.

Algorithm $A*$ is only an extension of the *Dijkstra* algorithm, where a heuristic is added to the cost calculation for each potential solution. This heuristic makes it possible to converge more quickly towards the optimal solution in those cases where we have prior information about this solution. For example, in the case of the shortest path in space, the heuristic can be Euclidean distance. So even if there are obstacles in the way, the algorithm will be attracted to the shortest distance possible. Therefore, the greatest difficulty for the A^* algorithm is to design a heuristic. One of the main criteria when designing this heuristic is that it represents the smallest minimum possible solution from a physical point of view. If this were not the case, the algorithm could converge towards the heuristic, and not towards the optimal solution (if this was lower than the heuristic).

In our application cases, the heuristic was determined by taking into account the physical distance between the starting point and the arrival point, and the minimum consumption over this distance if one were driving at constant speed. In this way, the heuristic results from a pre-optimization process, focusing only on travel time and energy consumption goals, at constant speed – something which is quickly solved by testing all the possible speed solutions and choosing the smallest one.

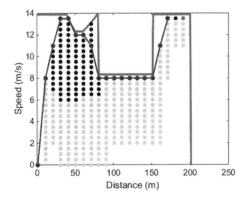

Figure 3.18. *Speed planning performed by an A* algorithm. In green, points open for evaluation; in black, evaluated points; in red, optimization constraints; in blue, optimum speed profile*

The results of the implementation of the A^* algorithm are shown in Figure 3.18. On this graph, we see that a small number of the points in the search space have been explored. In fact, space is constrained by the maximum physical or legal speeds, the nodes that we can explore are constrained by the physical limitations of the vehicle's dynamics (acceleration) as well as by psychological limitations – in terms of jerk – by drivers (or users). However, when the A^* algorithm is attracted by a heuristic, it does not necessarily explore all the existing solutions. As a consequence, the A^* algorithm makes calculations faster than *Dijkstra*, but may not find the optimal solution, if the heuristic is not correct.

These optimization methods have been used for electric motor vehicles as well. However, as these vehicles generally have the possibility of regenerating energy, *Dijkstra* proved unable to find a solution. Indeed, the regenerative properties of electric or hybrid traction chains can lead to negative loops because branches may have a negative cost (the regenerated energy corresponds to a negative cost). In this case, the algorithm remains blocked in this loop, and the cost of the optimal solution keeps decreasing. Although the *A** algorithm is also subjected to this theoretical limitation, it has proven to be capable of calculating solutions and been put in practice within the framework of the *eFuture* European project. However, this was only due to the particularities of the chosen heuristic, which made it possible to get out of the traps of negative loops.

3.3.1.2. Ant colony optimization

Ant colony optimization (ACO) has been used to fill the gaps in operational research methods, with the intention of calculating the vehicle's optimal speed profile for reducing fuel consumption, among other things. ACO seemed more promising for runtime considerations, as well as for avoiding the problem of negative loops. ACO can be used regardless of the vehicle's type of traction chain: electric, hybrid or thermal. In the way it was programmed here and contrary to graph searches in the graphs, this algorithm can only advance in the direction of the goal to be reached. ACO is therefore more flexible to implement in real applications. For that reason, linear scalarization has been chosen to achieve multi-objective planning. ACO (Colorni 1992) emulates the deposition and evaporation of pheromones on the paths of an ant colony in order to achieve the optimal speed profile, taking into account road conditions. ACO is a stochastic optimization method. At each stage, the ants choose the following speed with a probability expressed by a Bayesian filter:

$$
p_{ij}^k(t) = \begin{cases} \dfrac{\tau_{ij}^\alpha \cdot \eta_{ik}^\beta}{\sum \tau_{ij}^\alpha \cdot \eta_{ik}^\beta} & \forall j \in S \\ 0 & \forall j \notin S \end{cases}
\qquad [3.11]
$$

where τ is the sum of the remaining pheromones from previous ants and η is the cost to reach the new speed (cost function). τ can be considered as prior information in a Bayesian filter, whereas η can be thought of as the likelihood function in a Bayesian filter. S is the search space, i the current position, j the considered position and k the other positions which can be reached from the current position. In our implementation, the search space is broken down into square cells whose vertices are the nodes which can be explored by ants. The ant colony is launched from a chosen starting point, which can be the current speed of the vehicle, when the algorithm is launched. The advantage of ACO is that the destination point can be chosen much more freely. In our case, we have defined an arrival distance, but not a speed. The ant colony only needs to travel the distance from the starting point to complete the execution.

Good performance can be achieved by adjusting the main parameters of the ACO, that is to say, the chosen distance and the speed. Thus, for a distance of 50 meters, the algorithm operates in real time, whatever the speed.

The optimization cost function is calculated based on three cost functions: fuel consumption, travel time and driving comfort. Each part of the cost function is expressed as a function of vehicle speed. The ACO inputs are initial speed, legal speed, road inclination and traffic lights information. The ACO output is the optimal speed profile.

An example of the optimization using ACO is given in Figure 3.19 for an acceleration scenario. The red curve is the optimum speed towards which the ACO has converged. All blue lines are speed profiles that were temporarily considered as optimal during the optimization procedure. The maximum speed, shown in black, is defined as constant but may depend on road geometry and legal road speed, as in other planning methods. In this specific case, a high weight has been allocated to travel time.

This method was used during the European project *ecoDriver* in the eco-driving assistance smartphone application developed by IFSTTAR. This provided an indication of the speed to be kept by the driver in the form of a green zone on the speedometer. In the practical application of the algorithm, the ACO is rerun every 50 meters traveled.

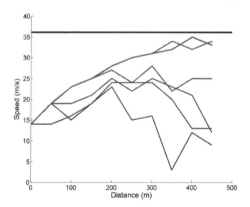

Figure 3.19. *Ant colony optimization: results on acceleration with a cost function set for maximum speed. Red shows the trajectory towards which the colony has converged. Blue shows other profiles proposed by the colony*

3.3.1.3. *MOSA (Multi-Objective Simulated Annealing) parametric method*

Here, a new method, aimed at being even more efficient in terms of computation time, has been explored in order to make it applicable for a real vehicle. This method is based on the simplification of the trajectory profiles. Instead of these being described by a series of points only constrained by the dynamics of the vehicle, they are functional profiles for which five parameters provide a sufficient description. The method is no longer discrete but continuous (piecewise), and space decomposition is no longer used. The disadvantage is that the route must be segmented into pieces, which no longer necessarily ensures the overall optimality of the solution. As a matter of fact, it is not possible to describe the whole of a trajectory using a single function. However, if the decomposition is done thoughtfully, it can altogether maintain the optimal character of the solution. Thus, setting up a space decomposition at each vehicle stop-point can preserve the validity of the optimality hypothesis. However, in this work, we have tried to go even further by describing the driver's behavior in a more detailed way in relation to reactions to speed change limitations on the infrastructure. Figure 3.20 shows two examples of theoretical speed profiles obtained from five parameters, as well as acceleration and deceleration functions resulting from the observation performed during the search cycle. In fact, a standard speed profile can be applied on each portion of a journey, built from the acceleration models by M. Treiber (2000). This piecewise breakdown assumes the existence of four types of speed profiles, depending on the aspect of speed limit changes.

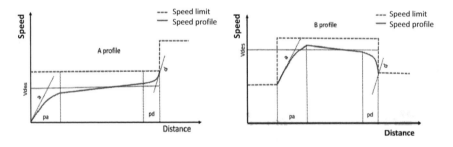

Figure 3.20. *Partial trajectory planning by functional representation and five parameters. a = acceleration parameter, d = deceleration parameter, Vdes = desired speed, Pa and Pd = portion of a journey where the a and d parameters are applied*

The results of this algorithm (first published in 2017 and then an extended version was released in 2019) (Figure 3.21 top) have shown the ability of the algorithm to calculate a set of trajectories in real time on a standard computation support. In this application, the optimal speed profiles are calculated beforehand over the entire journey of the vehicle. About 10 optimal speed profiles are calculated simultaneously in real time, which suggests a dynamic use of the algorithm. These results were then compared with standard trajectory planning methods, such as the *A** algorithm or the *Dijkstra* algorithm (Figure 3.21 bottom).

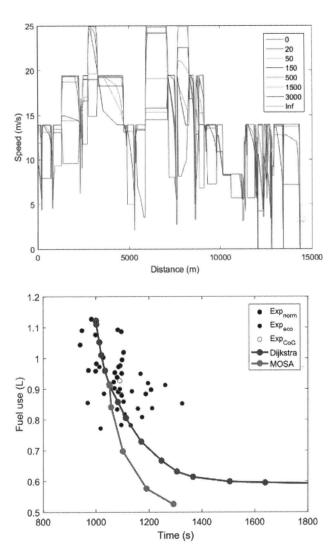

Figure 3.21. *Partial trajectory planning by functional representation and five parameters. Left: results obtained by piecewise optimization in real time (Orfila 2017, 2019); right: results in goals space and compared to A* operational research*

From the observation of the *Pareto* frontiers obtained, we can see that – under the simulated conditions – operational search algorithms such as *Dijkstra* make it possible to find more varied solutions than the proposed algorithm, but are more expensive in terms of computing time and less efficient than the MOSA algorithm. On this graph, the points represent trips made in real conditions on a route chosen by a set of 21 human drivers who have driven twice. We note that it is impossible for algorithms to predict optimal human behavior, since they have repeatedly exceeded the *Pareto* frontier. However, drivers who did not respect the Highway Code were not excluded from the analyses (speed favoring a better travel time, but not better energy consumption). This algorithm is currently being developed in the thesis by K. Hamdi (VEDECOM), where it is being evaluated in detail.

3.3.2. *Nonlinear scalarization*

The principle of nonlinear scalarization also involves determining a unique and global cost function. However, this is not a linear combination of the cost functions which depend on each of the goals. Here, the overall cost function is a nonlinear function of all the cost functions, and this overall cost function is defined by the algorithm's designer. In order to not favor one function over another, a certain expertise is required in the fields studied. For a global cost function f, we thus obtain the following equation:

$$f = g(f_i, \alpha_i) \hspace{3cm} [3.12]$$

This assumes that, in order to find the set of optimal solutions, the optimization process must be repeated by making α_i coefficients vary. In the following, an application of this strategy based on a genetic algorithm is presented.

3.3.2.1. *Genetic algorithm*

As nonlinear scalarization methods are very sensitive to designers' choices, some other techniques have started being envisaged since 2013. Thus, a first solution, detailed in the thesis of Olivier Orfila in 2009, then improved upon and presented in 2013, is based on the use of a genetic algorithm (Goldberg 1989) and plans both the longitudinal and lateral trajectory of the vehicle in having as objectives road safety, travel time and comfort. In addition, this algorithm reproduces human-like trajectory planning, the trajectory being permanently planned along the visibility distance, and updated according to a reaction distance, as shown by the algorithm below:

Algorithm 1 EWA Algorithm

Require: Initial conditions (speed, position)
Ensure: Optimized trajectory for complete itinerary
  while End of itinerary not reached **do**
    while $a_{max} \neq a_{min}$ **do**
      Optimize trajectory on D_{vis} to a_{max}
      Optimize trajectory on D_{vis} to a_{min}
      if $N_{a_{max}} > N_{a_{min}}$ **then**
        $a_{min} = a_{min} + ka_{min}$
      else
        $a_{max} = a_{max} - ka_{max}$
      end if
    end while
    Save trajectory start on D_{reac}
  end while
  Exit optimized trajectory

The steering wheel angle is optimized directly with the genetic algorithm. This algorithm generates a random initial population, then iterates a set of crossings, the probability of which depends on the evaluation of each solution, mutations and insertions of random individuals. For computation time reasons, speed is optimized by using a dichotomy. The overall cost function for each solution i is given by the following formula:

$$f(i) = \eta. \sum_{j=1}^{\eta_{genes}} \left[\left((x(j), y(j)) \epsilon R_0 \right) \wedge \left(a_y \leq a_{cond} \right) \right.$$
$$\left. \wedge \left(|\alpha_j| \leq \alpha_{max} \right) \right]. 100. \left(\eta_{genes} - j \right)^3 + (1 - \eta). T \qquad [3.13]$$

where j is the index of the individual's gene, η_{genes} the number of genes describing each individual, $(x (j), y (j))$ the spatial position in the coordinate system of the point corresponding to the gene studied, R_0 all the points belonging to the traffic lane, a_y the lateral acceleration of the vehicle, a_{cond} the acceleration limit supported by the drivers, α_j the vehicle's drift, α_{max} the vehicle's maximum drift, T the individual's travel time and η a coefficient between 0 and 1. Expression $\left[\left((x(j), y(j)) \epsilon R_0 \right) \wedge \right.$ $\left. \left(a_y \leq a_{cond} \right) \wedge \left(|\alpha_j| \leq \alpha_{max} \right) \right]$ is a Boolean which is equal to 1 when the three criteria – presence on the traffic lane, acceleration smaller than the driver's acceleration limit and vehicle drift smaller than the vehicle's maximum drift – are validated.

Figure 3.22 shows the results of this algorithm on a particularly demanding road for the vehicle's dynamics, with curvature radii smaller than 40 m, and a speed limit of 90 km/h. This work has shown the ability of the algorithm to calculate a real-time solution for speeds smaller than 70 km/h. It can therefore serve as a reference, but is currently difficult to use in real high speed conditions. When this algorithm operates, it looks like the movement of an earthworm, which gave it its name: *EarthWorm Algorithm* (EWA).

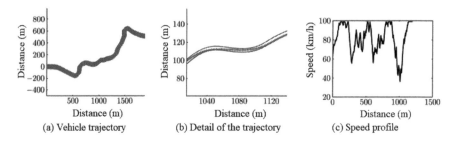

(a) Vehicle trajectory   (b) Detail of the trajectory   (c) Speed profile

Figure 3.22. *Multi-objective trajectory planning by nonlinear EWA scalarization*

A sensitivity analysis to the numerous parameters of the algorithm was carried out, and can be found in the thesis by Olivier Orfila (2009).

3.3.3. *Ideal methods*

Unlike scalarization methods, which seek one solution at a time, ideal methods seek to determine the *Pareto* frontier by making potential solutions interact with one another. The *Pareto* frontier is therefore a progressive convergence of the whole of solutions. Figure 3.23 shows the "*solutions*" convergence obtained by ideal methods.

There are many ideal methods, generally based on evolutionary algorithms, which natively let a population of solutions evolve towards an optimal population.

The basic principle of this type of ideal method is to evaluate the solutions not directly from their cost functions, but from a new measure of adaptation based on the strength and density of the solutions:

– Raw evaluation: the raw evaluation of the solution is determined by its dominant character. The more a solution dominates the others, the more important its raw evaluation will be. This raw evaluation is therefore dependent on the values of its cost functions.

– Diversity: the diversity of solutions is estimated from the concentration of solutions on the same part of the objective space. The farther the solutions are in terms of the goals, the weaker the diversity criterion will be.

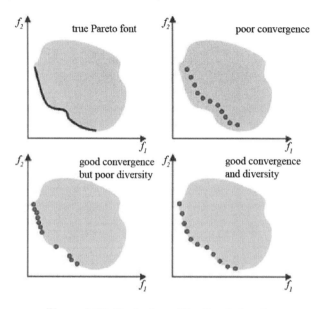

Figure 3.23. *Illustration of the Pareto frontier convergence and possible problems*

3.3.3.1. *Strength Pareto Evolutionary Algorithm 2 (SPEA2)*

The algorithm derived from SPEA2 (Zitzler 2001), developed in the works by Olivier Orfila, refers back to the nonlinear scalarization problem, that is to say, planning the steering wheel's angle and speed, but with a different multi-objective optimization method. The SPEA2 algorithm evaluates each individual in the population based on their strength and the density of surrounding solutions in the objective space. Thus, the cost function is defined by:

$$f(i) = R(i) + D(i) \tag{3.14}$$

where $f(i)$ is the cost function, $R(i)$ the raw adaptation depending on the dominant aspect of the solution, and $D(i)$ the evaluation of solution density.

The raw adaptation of the solution is evaluated by summing up the adaptation solutions which dominate it. This is a recursive algorithm which gives a null evaluation for a non-dominated solution, and the latter is calculated by using the following equation as a starting point:

$$R(i) = \sum_{j \in PUA, j \geq i} S(j)$$ [3.15]

where P is the entire population and A is an archive of the best solutions having appeared throughout generations of the genetic algorithm.

For each solution, the solution density is evaluated by classifying the distance between this solution and the other solutions. The kth furthest solution is the one chosen as an indicator. It is generally accepted that $k = \sqrt{|P \cup A|}$. Then, the density indicator $D(i)$ is calculated from:

$$D(i) = \frac{1}{\sigma_i^k + 2}$$ [3.16]

where σ_i^k is the distance in the objective space between the individual i and its kth nearest neighbor. As shown in Figure 3.24, the principle of the algorithm operates like the EWA algorithm, that is to say, it plans the trajectory along the visibility distance (corresponding to the driver's perception), and then keeps only the portion planned for the distance reaction (which corresponds to a time delay).

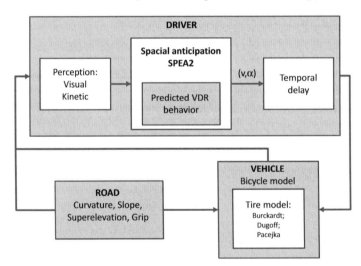

Figure 3.24. *Block diagram of the multi-objective planning algorithm based on the Strength Pareto Evolutionary Algorithm 2 (SPEA2)*

Figure 3.25 shows that during an algorithm iteration, the current *Pareto* frontier (in red) tries to converge towards the real *Pareto* frontier. Here, three objectives compete with one another (energy consumption, travel time and safety).

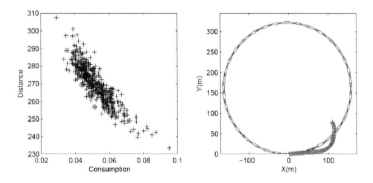

Figure 3.25. *Results of the algorithm derived from the Strength Pareto Evolutionary Algorithm 2 (SPEA2). Left: plot of solutions generated by the algorithm in the goals space (dominant solutions circled in red). Right: example of an evaluated solution*

3.3.4. *Summary of multi-objective planning methods*

During the development of the planning methods, a Java speed planning library was developed. This library contains the *A**, *Dijkstra*, *CO* and *genetic algorithm* methods. This library was used in the *ecoDriver* project, offline, to generate the optimal speed profiles. This library is interfaced with the *MapQuest* API (Open Street Map mirror database) and can therefore generate speed profiles using an origin and a destination directly selected on a map. Table 3.1 shows a summary of the different trajectory planning methods presented in this chapter. These cover a large number of different optimization methods, as well as the main multi-objective optimization methods (linear scalarization, nonlinear scalarization, ideal method).

3.3.5. *High level information*

As seen previously, multi-objective trajectory planning makes it possible to generate a set of optimal solutions by building a *Pareto* frontier. However, it does not make it possible to know which trajectory should actually be applied to the vehicle. In his book on multi-objective optimization, Deb (2009) partially addresses this question where he considers that it is absolutely necessary to use high level information in order to be able to choose which trajectory should be applied. This high-level information (see Figure 3.26) corresponds to knowledge which can be provided by the system's user, or the system's designer, or even by an external decision-maker.

Algorithm	Optimization Method	Optimization variables	Costfuncti ons	Multi-objectives	Comments
A* ICE (Internal Combustion Engine)	Operational search	Speed V(s) function of curvilinear abscissa	Travel time, energy use	Linear scalarization	Barely efficient decomposition by square cells
A* electric	Operational search	Speed V(s) function of curvilinear abscissa	Travel time, energy use	Linear scalarization	Space decomposition into square cells, negative loop effects not compensated by A* heuristics
Dijkstra	Operational search	Speed V(s) function of curvilinear abscissa	Travel time, energy use	Linear scalarization	Space decomposition into square cells, impossible in real-time (too long computation time)
ACO	Ant colony optimization	Speed V(s) function of curvilinear abscissa	Travel time, energy use, comfort	Linear scalarization	Space decomposition into square cells, fast calculation, highly sensitive to parameters (number of ants in the colony)
MOSA	Simulated annealing	Initial acceleration a, desired speed V_d, final deceleration	Travel time, energy use	Linear scalarization	Piecewise trajectory, parametric representation of solutions, sub-optimal structure
EWA	Genetic algorithm	Acceleration vs time $a(t)$, steering wheel angle vs time $\alpha(t)$	Safety, travel time, comfort	Non-linear scalarization	Vectorial representation, costly calculation
SPEA2	Genetic algorithm	Acceleration vs time $a(t)$, steering wheel angle vs time $\alpha(t)$	Travel time, energy use, comfort	Ideal method	Trajectory vectorial representation

Table 3.1. *Summary of multi-objective planning methods*

In scalarization methods, this information can be used before calculating optimal trajectory, but this may need recalculating if the high level information varies. In ideal methods, high-level information is generally used after calculating trajectories, which makes the set more dynamic in the event of a change in the high-level information, but requires longer computation time. In the context of Olivier Orfila's work on the problem of multi-objective planning for automated vehicles, several options for determining high-level information were considered:

– Manual tuning: in the form of a driving style selection that can be made by the driver. Evidently, this is only possible in the case of vehicles for individual use (private vehicles, individual taxis).

– Automatic learning of the path to be chosen: for shared-driving vehicles with self-driving mode, but where the user can regain control, it is possible to set up human driver learning preferences, by comparing the outputs of the multi-objective planning algorithm with the trajectory chosen by the human driver. However, if the trajectory is acceptable for the driver, nothing proves that it will be the chosen one when the driver becomes a passenger.

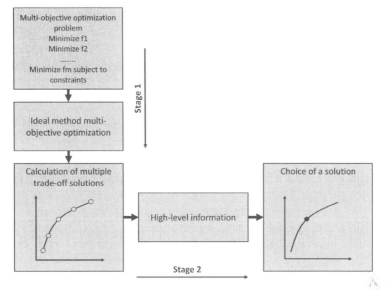

Figure 3.26. *Place of high-level information in multi-objective trajectory planning, according to Deb (2009)*

3.4. Conclusion on multi-agent planning for a fleet of vehicles: the future of planning

Building the trajectory of a vehicle in an isolated manner not only poses the problem of asymptotic stability which we mentioned at the beginning of this chapter, but also the problem of large-scale efficiency of autonomous vehicles designed in this way. It is therefore important to make an *a priori* evaluation of the impact of the strategies developed in isolation, as well as to design optimal systems for road networks. It would be interesting to design a planning for a multi-agent system, which takes into account multi-objective optimization considerations.

As much as bio-inspiration can be quite intuitive when it comes to developing a trajectory planner for an isolated vehicle, it is eminently complex when it comes to applying it to a fleet of vehicles. However, existing methods can be perfectly effective in dealing with these problems. These methods are often bio-inspired, as in the case of particle swarm optimization (method based on meta-heuristics). As there are many methods and multi-agent architectures, at IFSTTAR and in 2018, we launched two new research avenues (theses of J. Leroy and M. Tu) on how to supervise a fleet of automated and connected vehicles in mixed traffic conditions including conventional vehicles. These works are being tutored by Dominique Gruyer, Nour-Eddin El Faouzi and Olivier Orfila. The first results obtained in

Meiting Tu's thesis, jointly supervised with Tongji University (Shanghai, China), show the potential of vehicle supervision in a study for optimizing cab-sharing trips in the city of Chengdu, the provincial capital of Sichuan, in China. The company Didi Chuxing provided the data for 50,000 vehicles in one month, some of which have shared access. This made it possible to carry out a study on the impact of new supervision strategies, making it possible to increase the system's overall efficiency (either by reducing journeys or by maximizing user savings, thus maximizing shared journeys). As regards the second line of research, a generic meta-model is under development. This model can be broken down into different driving models: *"human"*, *"human and communicative"*, *"automated"* and, finally, *"automated and connected"*. These models applied in mixed traffic conditions will make it possible to approach the development of a high-level and more global strategy for traffic regulation by using CAVs as adjustment and optimization variables.

3.5. References

Bounini, F., Gingras, D., Pollart, H., Gruyer, D. (2017). Modified artificial potential field method for online path planning applications. *IEEE Intelligent Vehicles Symposium 2017 (IV 2017)*, June 11–14, Crown Plaza, Redondo Beach, CA, USA.

Claussmann, L. (2019). Motion planning for autonomous highway driving: A unified architecture for decision-maker and trajectory generator. PhD Thesis, University of Paris, Saclay, France.

Colorni, A., Dorigo, M., Maniezzo, V., Varela, F.J., Bourgine, P. (1992). Distributed optimization by ant colonies. *Proceedings of ECAL91 – European Conference on Artificial Life*, Paris, France.

Deb, K. (2009). *Multi-objective Optimization Using Evolutionary Algorithms*. Indian Institute of Technology, Kanpur, India.

Dijkstra, E.W. (1959). A note on two problems in connexion with graphs. *Numerische Mathematik*, 1, 269–271.

Goldberg, D.E. (1989). *Genetic Algorithms in Search, Optimization and Machine Learning*, 1st edition. Addison-Wesley Longman Publishing Co., Inc., USA.

Gruyer, D., Cord, A., Belaroussi, R. (2013). Vehicle detection and tracking by collaborative fusion between laser scanner and camera. *IEEE/RSJ International Conference on Intelligent Robots and Systems IROS'13*.

Hart, P.E., Nilsson, N.J., Raphael, B. (1968). A formal basis for the heuristic determination of minimum cost paths. *IEEE Transactions on Systems Science and Cybernetics*, 4(2), 100–107.

Jouandet, M. and Gazzaniga, M.S. (1979). The frontal lobes. In *Handbook of Behavioral Neurobiology, Volume 2*, Gazzaniga, M.S. (ed.). Plenum, New York, USA.

LaValle, S.M. (2006). *Planning Algorithms.* Cambridge University Press, Cambridge, UK.

Michon, J.A. (1979). *Routeplanning en geleiding: Een literatuurstudie.* Institute for Perception TNO, Soesterberg, The Netherlands.

Orfila, O. (2009). Influence de l'infrastructure routière sur l'occurrence des pertesde contrôle de véhicules légers en virage : modélisation et validation sur siteexpérimental. Mémoire de thèse, Université d'Evry, France.

Orfila, O. (2011). Impact of the penetration rate of ecodriving on fuel consumption and traffic congestion. *YR*, Copenhagen, Denmark.

Owen, A.M. (1997). Cognitive planning in humans: Neuropsychological, neuroanatomical and neuropharmacological perspectives. *Progress in Neurobiology*, 53, 431–450.

Rajamani, R. (2006). *Vehicle Dynamics and Control.* Springer, Cham, Switzerland.

Revilloud, M., Gruyer, D., Pollard, E. (2013). A new approach for robust road marking detection and tracking applied to multi-lane estimation. *IEEE IV 2013*, Gold Coast, Australia.

Sattel, T. and Brandt, T. (2008). From robotics to automotive: Lane-keeping and collision avoidance based on elastic bands. *Vehicle System Dynamics*, 46(7), 597–619.

To, T., Meinecke, M., Schroven, F., Nedevschi, S., Knaup, J. (2008). CityACC – On the way towards an intelligent autonomous driving. *Proceedings of World Congress of the International Federation of Automatic Control*, 6–11.

Treiber, M., Hennecke, A., Helbing, S. (2000). Congested traffic states in empirical observations and microscopic simulations. *Physical Review E*, DOI:10.1103/PhysRevE. 62.1805.

Vanholme, B. (2012). Highly automated driving on highways based on legal safety. PhD Thesis, Université d'Evry-Val d'Essonne, France.

Vanholme, B., Lusetti, B., Gruyer, D., Glaser, S., Mammar, S. (2011). A highly autonomous driving system on automotive microprocessors. *IEEE ITSC 2011*, October 5–7, The Georges Washington University, Washington D.C., USA.

Wolf, M.T. and Burdick, J.W. (2008). Artificial potential functions for highway driving with collision avoidance. *IEEE International Conference on Robotics and Automation (ICRA)*, pp. 29, 30 and 46.

Zitzler, E. (2001). Spea2: Improving the performance of the strength pareto evolutionary algorithm. Paper, Computer Engineering and Communication Networks Lab (TIK), Swiss Federal Institute of Technology (ETH) Zurich, Switzerland.

4

From Virtual to Real, How to Prototype, Test, Evaluate and Validate ADAS for the Automated and Connected Vehicle?

4.1. Context and goals

Over the last decade, the rapid development of vehicle information systems and embedded sensor technologies has highlighted a growing need for automotive manufacturers, as well as for research laboratories, to find the means to prototype, test, and evaluate complex embedded systems (ADAS: *Advance Driving Assistance Systems*) which can be active and cooperative. Since 2014, EuroNCAP has integrated several driver assistance systems during their homologation process, which are grouped under the title "*Safety Assist*" and weigh 20% of the final rating. Among these systems we can mention AEB (*Automatic Emergency Braking*), for example.

However, the testing and validation process is either limited to a small number of use cases (three for AEB), or only to works with data provided by car manufacturers (this is the case for lane-keeping applications). However, road safety and the risk of a situation are directly related to the reliability and robustness of these embedded systems, and to the information retrieved by them. In recent years, we have also seen that the sensors needed for the use of these so-called "*intelligent*" driving assistance systems are increasingly ubiquitous, numerous and complex. In order to evaluate the performance and quality of applications and algorithms for information processing, decision-making, and control/command (included in the design of automation

Chapter written by Dominique Gruyer, Serge Laverdure, Jean-Sébastien Berthy, Philippe Desouza and Mokrane Hadj-Bachir.
For a color version of all the figures in this chapter, see www.iste.co.uk/bensrhair/adas.zip.

systems for partial, full, and/or shared driving), it is necessary to develop procedures, measurement tools, and ground truths. In view of the diversity of the situations to be tested (taking into account possible degradations and adverse conditions: climate, infrastructure, sensors, etc.), it is increasingly difficult to perform these tests only on test tracks. Consequently, alternative and, above all, complementary solutions must be found in order to be able to take into consideration a large number of situations and data. For this, the use of testing and simulation tools and platforms is becoming essential. In addition, it is necessary to be able to interface all these testing and simulation tools in order to obtain exploitable and, most of all, valid results. Of course, taking the human aspect into account for these tools makes the problem more complex and requires real-time operation, something which can be as close as possible to reality. With a pre-certification in view, the evaluation of active and cooperative ADAS or AdCoS (*Adaptive Cooperative Human-Machine Systems*) is a major issue for automated driving applications.

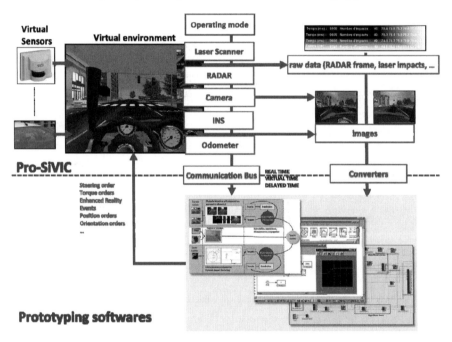

Figure 4.1. *Virtuality, an alternative solution for the prototyping and testing of complex and looped automated mobility systems*

In fact, when compared to more traditional ADAS, the approach to designing and evaluating AdCoS presents specific difficulties inherent in the "*adaptive*" nature of these systems: while their strength lies in their ability to adapt to the driving context in real time, in order to better cooperate with humans and meet their actual needs, their ability to adapt is also a source of infinite variability, which in turn poses real challenges in terms of evaluation and validation. Validating an AdCoS presupposes being able to test it "*in interaction with (simulated or real) humans*", in order to understand its functioning and its efficiency, acknowledging a diversity of situations, practices and potential uses that will be made of it by the final driver, something which involves the disengagement from the automated system and its taking over by the person in charge of the driving task, in thorough safety conditions.

The challenge is no longer just "*technological*" certification as such, but also certifying this technology "*from the perspective of human use*", which is itself adaptive and changing. Currently, there are no generic design processes, no standardized methods, and almost no integrated simulation tools to perform such tests for the design, development and evaluation of AdCoS. This is also true for the evaluation of active AdCoS interaction in a fleet of vehicles fully or partially equipped with ADAS and/or means of communication. The Pro-SiVIC simulation platform as we propose it undoubtedly provides a first alternative and effective solution by:

– decreasing the number of kilometers required to travel in the real world to evaluate a system, or a "*system of systems*", in a large and rich set of critical scenarios, which can involve infrastructure (roadside) degradations, climatic conditions, sensors, and algorithms;

– ensuring the repeatability and reproducibility of driving conditions and equipment;

– measuring the performance of ADAS, PADAS (*Partially Autonomous Driving Assistance Systems*) and AdCoS with a reliable and precise generation of "*ground truths*";

– integrating human and cognitive engineering aspects, including monitoring considerations, Human-Machine Cooperation, and simulation practices on how end drivers could use these AdCoS.

In this chapter, we will present this alternative upstream simulation solution, so as to address the above-listed problems. This solution is based on the use of the Pro-SiVIC platform. This interoperable, modular, and dynamic platform makes it possible to respond perfectly and efficiently to the constraints imposed during the implementation of an ADAS evaluation and validation process. This chapter will begin by presenting the general architecture, and then continue with the enumeration of the main functionalities it is necessary to implement in such a simulation platform in order to guarantee a high level of representativeness of the simulated environments and of the modeled sensors. In addition, some application examples will be presented. These examples will highlight the effectiveness of simulation for prototyping, testing, and evaluating active and cooperative ADAS.

4.2. Generic dynamic and distributed architecture

4.2.1. *Introduction*

The simulation of (partial or complete) connected automated vehicle systems requires the implementation of an assembly of several software tools and hardware peripherals.

This architecture will be dependent on the richness of the environment where our system is placed for validation/evaluation purposes.

Indeed, several modules come into consideration in order to create and enrich this environment:

– traffic simulation models and applications;

– vehicle dynamics simulation models and applications;

– models and modules of vehicle embedded sensors;

– data processing and control/command algorithms;

– models and communication module for C-ITS;

– a simulation engine enabling the visual rendering, orchestration/ synchronization of the various simulators and models, event management, and resource management (graphics, *Surface Equivalent Radar* or SER, *Bidirectional Reflectance Distribution Function* or BRDF, etc).

In addition to these modules being interconnected with one another, other needs must be taken into account by Pro-SiVIC and, by extension, by simulation platforms, such as the ability to interact with a human being in the loop: a driver, a passenger or as a key component in the road scene (pedestrian, cyclist, other user, etc.).

In order to interact with humans, there is a need to add a certain number of peripherals to the platform that will enable them to be immersed in the scene and interact with it. Depending on the quality of the desired immersion, visual rendering and preferred feedback, a wide range of equipment must be able to be connected to the simulation platform. This can range from a simple steering wheel and pedals, to dynamic simulators for reproducing movements which are representative of the vehicle's dynamics. More and more, interaction with humans is requiring the use of immersive (virtual reality helmet) and haptic means (force feedback systems, vibrating systems, etc.).

The last consideration, which will constrain the choice of architecture for the construction of a simulation platform, is its ability to run *"real-time"* simulations. However, depending on the representativeness, richness and complexity of certain models (vehicle dynamics, sensor models, etc.), it may be necessary to distribute the calculations along several processor cores, or across several computers. This distribution of the calculations will make it possible to efficiently accelerate the execution of calculations or models.

4.2.2. *An interoperable platform*

As we mentioned in the introduction, the need to interconnect different tools and models with one another has become a crucial need. It is with this intention that a general standard with the acronym FMI (*Functional Mockup Interface*) was created, for easing up the exchange of models and standardizing the way of connecting and sequencing them.

The FMI was first deployed to bring together multi-physical and heterogeneous models running on different tools. The standard has been progressively enriched to support use cases which are specific to the simulation of a perception system, by supporting the transfer of large amounts of information (video streaming, for example).

The other interest of the FMI is its protocol, which makes it possible to manage co-simulation by implementing different tools, based on *"Master-Slave"* logic. The goal is to provide a standard interface to couple several tools in a co-simulation environment. The data exchanged between the sub-systems is limited to a discrete communication point. In the time interval between two communication points, the sub-systems are solved independently from one another, thanks to their own solver or simulator. *"Master"* algorithms control the data exchange between the sub-systems and the synchronization of all *"slave"* simulation solvers.

The FMI is a protocol which makes it possible to assemble relevant multi-physical models, such as the dynamic behavior of a mobile (car, motorcycle, etc.), but is not suitable for asynchronous information sharing, as the sensor output for algorithmic processing.

To this end, the DDS (Data Distribution Service) data exchange protocol is currently used. This protocol is now widely supported and natively found in many development environments (the ROS type, for example).

DDS introduces a virtual global data space where applications can share information by simply reading and writing data "*objects*", addressed by using an application-defined name and a key. An example of distributed architecture is shown in Figure 4.2.

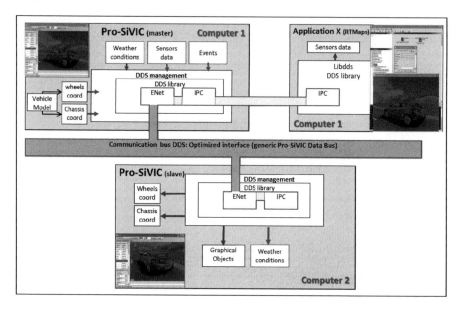

Figure 4.2. *DDS, an efficient communication solution for the distribution of resources and processing stages*

This protocol offers precise and extensive Quality of Service (*QoS*) control, including reliability, bandwidth, delivery times and resource limitations.

Implementations of this protocol, based on different TCP/IP – UDP/IP transport layers, make it possible to ensure a distributed implementation mixing different operating systems.

Figure 4.3 illustrates the different interconnection needs between the different types of modules for which the FMI and the DDS are candidate solutions.

In the same vein, for promoting open and interoperable simulation platforms, a recent initiative led by major manufacturers proposes to standardize sensor model interfaces.

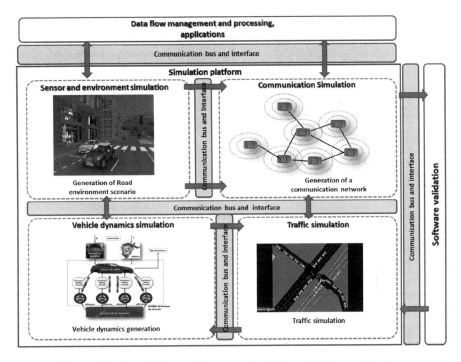

Figure 4.3. *Pro-SiVIC: an interoperable platform for the prototyping, testing, and evaluation of ADAS involving third party platforms*

This initiative is now supported by the ASAM (*Association for Standardization of Automation and Measuring systems*) consortium and is known by the acronym OSI (*Open Simulation Interface*). The goal of the OSI is to standardize data exchanges between the simulation environment, sensor simulation and function simulation tools.

So far, we have considered interoperability from the point of view of protocol and data and information exchange format, but there is also a standardization level which makes it possible to share the same understanding of a scene, its content, its road network and, more recently, a first level of scenario description.

The advantage of having a common description is that one can run the same simulation on different types of simulation platform, just by capitalizing on the latter.

4.3. Environment and climatic conditions

4.3.1. *Introduction*

Before being able to simulate the sensors, it is necessary to generate a rendering of the environment and the road scene, both in the visible electromagnetic domain and in a large set of wavelengths. In order to enable a physico-realistic simulation, it is imperative to generate and manage events, as well as a highly dynamic virtual world (spatially and temporally). For this aspect of object simulation, a dynamic loading library is required. This library can be considered as part of the core of the simulation engine and makes it possible to load and delete objects while the simulation is being executed. This highly modular and adjustable architecture is based on the use of *"plug-ins"*, which are dynamically loaded when the simulation engine is launched. In its initial version, the simulation engine was designed as an optimized graphics engine, with a set of the required adaptations to interpret meta-information (BRDF, SER) and dedicated mechanisms: *"multi-rendering"* manages several levels of rendering in terms of quality and wavelength (suitable for the modeling of particular sensors: GPS, RADAR, IR, etc.), *"post-processing filters"* enable processing and modifications applied to an initial rendering, and the *"management of rendering layers"* (visibility layer) helps manage the working view, the generation of constrained renderings (particular visible or invisible objects, metadata), and optical ground truths. In the case of complex scenes requiring the management of millions of facets (if we want to maintain real-time processing), it is necessary to optimize the management of graphic resources. A solution has been proposed using *BSP-trees* (binary space partitioning trees), which guarantee rapid and efficient graphics rendering. A set of additional generic classes and functions also makes it possible to manage graphic objects without optimization (*meshes*), textures (static, animated, procedural, etc.), materials and meta-materials, light sources, shadows, ray-tracing to manage collisions and physical interactions, control devices for objects of the scene, and the positioning properties of *"positionable"* objects. Some of these mechanisms are discussed and briefly presented in the following sections.

4.3.2. *Environmental modeling: lights, shadows, materials and textures*

In a graphics engine, in addition to the quality of the objects' mesh and the choice of materials and textures, the quality and realism of the rendering will be strongly impacted by the management level of light sources and the simulation of shadows. As it is possible to observe in everyday life, the mechanisms underlying shadow formation are multiple, and often cause objects to interact with one another. The first cause for the production of a shadow is direct and corresponds to the occultation of a space area by an object. It is the projected shadow that we can observe in Figure 4.4. However, the use of this type of shadow alone is not sufficient and causes a floating effect above the ground of the object casting the shadow. In order to remove this optical illusion, it is necessary to model and generate occluded shadows. During the graphic rendering of an object, this mechanism takes into account the occlusion of ambient light by other objects. The implemented mechanism samples the occlusion caused by each object following a small three-dimensional grid stored on the GPU, in the form of a 3D texture. The use of a small resolution is sufficient because of the low frequency nature of the ambient occlusion information. The actual calculation of ambient occlusion is performed by ray tracing. This calculation is only performed once for each polygonal model and its result is later stored on the hard drive. Apart from the moment when it is generated, this shadow, once calculated, does not consume any particular resource. The result of its application can be seen in Figure 4.4. Finally, we also implemented cast shadows (shadows cast from one object onto another object), and auto-shading which lets an object produce shadows on itself. The use of all these mechanisms produces a consistent rendering in terms of shading.

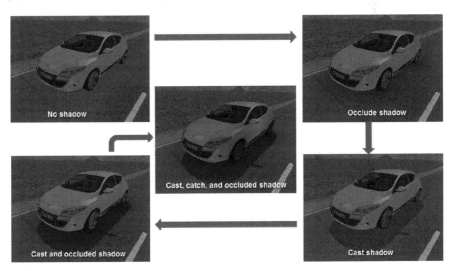

Figure 4.4. *Taking into account the different types of shadows: catch, cast, self-shading, occluded*

However, in order to be able to produce these shadows, it is necessary to know how to generate light sources. In addition to using an ambient light source, point sources can be added. For each of these sources, a set of parameters are accessible and configurable: position, orientation, colorimetric components, and environmental behavior (type of attenuation of the power of the light source as a function of distance). This attenuation uses a second degree polynomial formulation, which is a function of the distance and three configuration coefficients.

The consideration of light masks has been added to this mechanism. These masks will make it possible to simulate not only diurnal, but also nocturnal situations. Under these conditions, one can simulate the evolution of a vehicle with its headlights on. The rendering obtained by the use of these light masks can be observed in Figure 4.5. Each light mask corresponds to a texture modulating the intensity of the light in the emission zone of the light sources. In the example given, the masks encode a circular three-dimensional emission zone, which produces a volume projection of the light intensity.

In Figure 4.5, it is possible to observe the improvements made to this mechanism with a new physically realistic "*pixelated*" plug-in render. In these renderings, the headlights use realistic optical block illumination maps.

Figure 4.5. *Light masks and night lights (2 headlights, HDR texture, and reflection)*

In order to be able to generate a graphic rendering with a *High Dynamic Range* (or HDR), a mechanism for taking into account HDR textures has been implemented. This makes it possible to simulate the dynamics of optical type sensors in a more realistic way in the case of scenarios where the overall brightness of the images perceived by the virtual cameras can greatly vary over time (passage under a tunnel, coming across other vehicles at night, etc.). In order to be able to obtain an HDR rendering, two major operations must be applied: firstly, we must use a floating number render buffer suitable for HDR rendering (8 bits for normal rendering, 16 bits and 32 bits for HDR rendering), then the view has to be enhanced with a post-processing filter which performs the energy level conversion along $[0; +\infty]$ in brightness along the interval $[0; 1]$. This is a *Tone Mapping* global filter which maps a window of fixed energy values over the entire image in direction of the output interval, using a lookup table stored in a texture. This corresponds to the behavior of CCD, CMOS or movie cameras.

If one wishes to simulate the auto-exposure mechanism which controls the dynamic range perceived by an optical type sensor, then a new auto-exposure filter must be applied. However, with these techniques it is not possible to simulate human vision.

In order to simulate certain climatic and reflection effects, the structure of the simulation engine had to be revamped and adapted. Thanks to these modifications, it became possible to render dynamic textures during the final rendering of a scene. Generating these textures should only be done once per image, before rendering all the views. This mainly resulted in the creation of an abstract pre-rendering class used by all classes requiring a pre-processing before the final rendering.

Taking advantage of this change, a set of texture generation plugins was developed to simulate flat reflective surfaces or cubic reflective surfaces, for instance.

A first plug-in proposes to apply a filter to the *"average"* reflection on a rectangular box, in order to simulate the anisotropic shine of materials. Thus, we can simulate the reflections on a wet or damp road, using a box of significant height but small width: the reflections tend to stretch in height on a road due to bitumen roughness. With this plug-in one can also darken the reflection and apply a gamma correction, in order to reflect only high light energies. This mechanism is very interesting for simulating the reflections of objects on wet roads in night conditions, for example. Figure 4.6 shows two graphic renderings obtained with this plug-in. In this case, the reflection texture is modulated (at the material level) by a texture defining a reflectivity map of the scene, and by a second texture simulating the irregularities of the asphalt (bitumen and aggregates). The result of this combination of textures is related to the emissive component of *"road"* materials, which makes it possible to give a wet aspect to this road.

The second plug-in proposes to render the environment encompassing the object displayed in a *"cube-map"*. Color correction and blurring can also be applied to the generated texture, so this plug-in can be used both to calculate dynamic reflections – for example on vehicle bodies (Figure 4.7) – as well as to simulate indirect scene lighting over the object.

Figure 4.6. *Planar reflection mechanism to simulate a wet road*

Figure 4.7. *Environment reflection mechanism on an object for a more realistic rendering*

4.3.3. *Degraded, adverse and climatic conditions*

In order to make road scenes more realistic for "*all kinds of weather*" conditions, it is imperative to be able to simulate the climatic disturbances that one may come across in real situations. To this purpose, a library of post-processing filters has been produced, using the API available in the simulation engine. In addition to simulating climatic conditions, this library also offers the filters needed to modify the initial rendering of a camera and obtain a final rendering of a scene as close as possible to the physical behavior of an optical sensor. The application of these filters is achieved by using the "*multi-rendering*" mechanism.

The main filters currently available in this library are shown in Figure 4.8. This library brings together filters for optical, light exposure, and climate post-processing. A large majority of these filters have been developed using the "*shaders*" functionality, in order to be able to optimize calculation times.

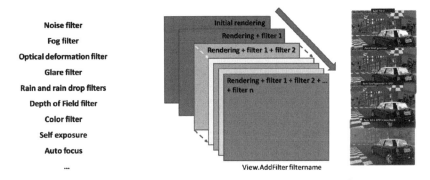

Noise filter

Fog filter

Optical deformation filter

Glare filter

Rain and rain drop filters

Depth of Field filter

Color filter

Self exposure

Auto focus

...

Initial rendering

Rendering + filter 1

Rendering + filter 1 + filter 2

Rendering + filter 1 + filter 2 + ... + filter n

View.AddFilter filtername

Figure 4.8. *"Multi-rendering" mechanism and post-processing filters*

4.3.3.1. *Rain simulation (drop on the camera lens, falling drops)*

To simulate the fall of raindrops, preliminary rain rendering tests based on textured vertical planes were performed, which notably led to the improvement of the *waterfall* procedural texture generation plug-in. The result was good for a fixed camera, but not satisfactory for a mobile camera, in particular because of the discontinuities observed when the camera crosses a "*rain plan*".

The approach adopted provides a simulation and rendering of each of the raindrop trajectories. To simulate falling rain, a chosen number of rain particles are evenly distributed within a configurable interval. When a drop of water appears, its lifespan is calculated by performing ray tracing, taking into account the falling speed vector. This ray tracing is not performed for each image, but retains a reasonable impact on performance, even for a large number of particles. When a rain particle exceeds its allotted lifespan, it is destroyed and replaced with a new particle. Particles interact with all renderable objects in an image. In order to properly simulate *motion blur*, the render takes camera movement into account, and distorts the trajectories displayed as a function of the former. The opacity of the particles can also be controlled in an inversely proportional manner to the distance traveled in relation to the camera.

This type of rendering ensures a realistic result in the event of rapid camera movement (turning at an intersection, for example). The rain rendering plug-in also has accessors which can be controlled by a third-party plug-in (dynamic control of the rain intensity, etc.). This plug-in also makes it possible to take into account multiple cameras, with different refresh rates. Figure 4.9 shows the rendering obtained for this rain filter. It also shows how environmental objects are taken into account in the evolution of water drops. Thus, no drop is present under the bridge. In Figure 4.9, we can see that by adapting the configuration of the raindrops (width, speed, opacity, color), we can obtain a snowfall type rendering. This figure also shows the impact of the observer's movement on particle dynamics.

A second rain post-processing filter was developed to simulate the distortion of the image caused locally by drops of water on the lens of a camera, or on the windshield of the vehicle in the case of an embedded camera in the car cabin. This deformation is based on a texture whose luminance represents the thickness of the film of water at each point of the image.

Figure 4.9. *Rendering of a snowfall and impact of observer dynamics on the evolution of raindrops*

This texture can optionally be a motionless image, an animation or even a procedural texture. The raindrop dynamic texture generation plug-in was designed as a "*companion*" for the rain filter.

This texture generation plug-in can procedurally create an animation corresponding to water drops falling on a plane. Its guiding parameters are texture resolution, minimum and maximum drop size, the number of drops appearing per second, and the fading factor, which represents the speed with which the drops flow and disappear. For the rain filter, its parameters are the maximum thickness of the water film and the refractive index of the water film. Figure 4.10 shows the real-time rendering of water drops on the lens of a camera, and the effect of high water density on the lens.

Figure 4.10. *Raindrop rendering on the lens of a camera*

4.3.3.2. Fog simulation

The second climatic effect degrading the visibility of a scene and having a significant impact on the images produced by the optical sensors is fog. A new plug-in has been developed to add a fog effect to the rendering, *a posteriori*. The result is more realistic than using OpenGL fog and follows *Koschmieder* law. Thus, for each pixel, the color obtained at the output of the filter depends on the color of the input pixel, the distance of the pixel from the camera, the fog density and the fog brightness. The setting of this filter is made according to the fog density, the brightness of the sky, the type of attenuation (constant, linear, or quadratic), and the fog color. One can also fix a fog zone depending on its position and its radius of application. In order to obtain inhomogeneous fog, it is necessary to apply a post-processing on the clear image (without fog). This inhomogeneous fog simulation is unfortunately limited to static cameras. If one wishes to extend it to moving cameras, it is necessary to manage volume fog.

Figure 4.11 shows the generation of these two types of fog (homogeneous and inhomogeneous). This scene of an urban environment with fog was used extensively in the context of a thesis dedicated to image restoration (Halmaoui 2013; Tarel 2010, 2012). A database named FRIDA (Foggy Road Image DAtabase) was generated with Pro-SiVIC platform and is available for the scientific community at the following address: http://perso.lcpc.fr/tarel.jean-philippe/bdd/frida.html.

Of course, all these weather condition filters can be activated and used at the same time. However, one must take into account that the order in which they are set corresponds to the order in which they will be applied.

Figure 4.11. *Homogeneous and inhomogeneous rendering of daytime fog*

4.3.4. *Visibility layers and ground truths*

When several views are handled at the same time and it is necessary to display particular information specific for each of them, it is important to manage a mechanism for the visibility level of the objects. A function for managing visibility layers called *"layers"* has been added to the simulation engine. This mechanism meets the following set of constraints: availability of a working view, *"bounding boxes"* management, and generation of ground truths.

Thanks to this functionality, in Pro-SiVIC it is possible to very easily configure, during simulation, the visibility of the scene's elements for each particular view. For example, in Figure 4.12, several views are manipulated with a particular purpose for each. A first view corresponds to the simulation of an optical sensor, whereas the other views simply offer the user working and control windows (viewpoint positions and light sources). In Figure 4.12, some views have been enriched with a rendering of additional information materializing certain objects which are usually invisible: light sources, sensors, interest points, etc.

In this mechanism, the visibility of the elements is managed according to layer logic: each displayable entity of the scene is present or not on a certain number of layers, and each view displays only the objects which are present on at least one layer within a selection. The set of layers on which an object is present is represented by a positive integer. Each layer corresponds to a power of 2 (1, 2, 4, 8, 16, 32, and so on). The set of layers on which an object is present is represented by the sum of the values corresponding to each layer. Thus, an object present on the first and the third layer reunites the value of layers: $1 + 4 = 5$. The layer value of an object can be modified in real time by changing its *"layers"* property. If one does not define this property, objects are present on all layers. 24 visibility layers are available ($0 \leq$ layers $\leq 2^{24}$-1).

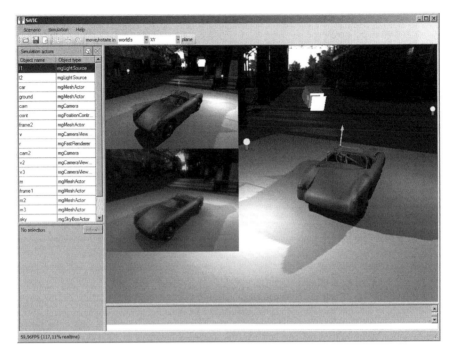

Figure 4.12. *Use of visibility layers for the management of a working and a control window*

Besides this working window functionality, this mechanism offers a powerful tool to generate "*ground truths*" for optical processing. Indeed, for the validation of the obstacle detection and road marking detection algorithms, it is possible, simultaneously with image generation, to obtain labeling images (of the obstacles and markings).

In addition, for the simulation of a GPS type sensor, this mechanism is used so as to generate bounding boxes modeling the 3D objects in the environment. This simplified modeling of objects can then be used either to manage collisions, or to simulate the effects of GPS signal multi-reflections coming from satellites. Figure 4.13 shows the result of a simulation of a GPS sensor using this "*bounding boxes*" mechanism.

In Figure 4.14, the use of the *z-buffer* provides a depth map and the "*layer*" for the application of a specific material, which is only visible in this render *layer*, making it possible to generate a reference containing the pixels of an object. This segmented image can be used for quantifying the quality of obstacle detection by an optical sensor (visible or thermal IR). Figure 4.14 shows an example of stereovision processing (dense disparity map, vehicle extraction and tracking), as well as the generated ground truths (depth map and vehicle labeling).

Figure 4.13. *Visibility layer for collision and multi-reflection management*

Figure 4.14. *Vehicle labeling and generation of a depth map*

Figure 4.15 presents the generation of a ground truth with a marking mask enabling access to the pixels which exclusively constitute the marking. This figure is obtained by using a physico-realistic simulation of Satory's test tracks (Versailles, France). Regarding ground truths, excluding optical rendering, for static and dynamic objects, a reference sensor is available ("*observer*"). This "*observer*" will produce a state vector containing the current state of the object.

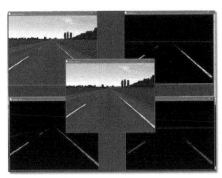

Figure 4.15. *Labeling of road markings*

4.4. Modeling of perception sensors

The functions and mechanisms presented so far already make it possible to have an interoperable rendering and simulation architecture which is sufficiently generic so as to be interconnected with a set of third-party applications, enabling the generation of complex and critical scenarios. The next step is to put in place the mechanisms and models that will generate the data required for ADAS prototyping. To achieve this task, the capabilities and functionalities of the simulation engine will be used for simulating sensor operation. The purpose is to generate the data produced by the sensors in both embedded and off-board modes. In order to remain consistent with our initial philosophy and keep the advantages associated with the use of plug-ins and dynamic class loading, an additional library of sensors will be proposed. These sensors, which are plug-ins, are both proprioceptive (Odometer, Inertial Navigation System) and exteroceptive (conventional, omnidirectional and Fish-eye cameras, laser scanner, RADARs) (Gruyer 2010, 2012; Pechberti 2012, 2013, 2018; Hadj-Bachir 2019a), or communicating sensors (Transponder and 802.11p communication) (Demmel 2014, Gruyer 2013). In its current state, this library offers the majority of sensors to be used for the development of ADAS, either in an embedded configuration (for dynamic objects such as cars, trucks, buses, shuttles, pedestrians, etc.) or off-board configuration (positioned on the infrastructure/road side).

4.4.1. *Typology of sensor technologies*

Perception from a vehicle is a crucial step for designing active driving assistance systems. Indeed, trajectory planning and "automated" driving strategies are directly dependent on the quality and exhaustive character of the perception stage. This perception of the environment can be broken down into several stages. The first one is related to the sensors and the data they will produce. The second one concerns data filtering and (spatial or temporal) alignment. The third stage involves segmenting data and extracting the most synthetic and exploitable objects and knowledge. The fourth stage aims to extract and generate a semantic and dynamic layer in order to gain a better understanding of the events in the environment close to the vehicle. This knowledge will make it possible to predict and anticipate future situations. The outputs of stages 3 and 4 will enable the vehicle to have a representation of the environment (Local Dynamic Perception Map) for developing planning strategies (path, trajectory, control) and making decisions.

This Local Dynamic Perception Map will contain a "*picture*" of the environment with an estimate of the road scene "*key components*" (obstacles, road, ego-vehicle, environment, and driver). For example, for the "obstacle" key component, its attributes could be the nature of the object, its positioning (coordinates), its displacement speed, its attitude or its temporal behavior.

To acquire data which is representative of the environment, a set of sensors (with their intrinsic and extrinsic characteristics) is available. Each sensor category will have its own operating domain with different frequency bands (physical and technological domains). In addition, each of these sensors is not able to perceive the exhaustive character of a scene in "*all kind of weather*" (environmental) conditions. Each sensor will have operating limitations and therefore be affected by specific conditions of use (illumination, fog, rain, etc.).

To compensate for these limitations and these perceptual weaknesses (which are critical at the perception stage), it is essential to be able to use several types of sensors, resorting to different kinds of technologies, at the same time. The plurality of perception sources will make it possible to improve the quality, accuracy, reliability, robustness and richness of perception. The purpose is to exploit the redundancy and complementarity of sensors and information in order to guarantee a high level of perception quality.

The main sensors used for the development of ADAS, PADAS, and AdCoS are:

– optical sensors (CCD and CMOS camera, fish-eye camera, omnidirectional camera, etc.);

– LiDAR sensors (scanning, solid state);

– RADAR sensors (long range, middle range, short range);

– GNSS sensors (natural, differential, RTK (real-time kinematic)).

4.4.2. *From a functional model to a physical model*

The validation of the functions used in an automated and connected driving application is based on the validation of the sub-systems which make up this function, and their combined operation. The validation of the various sub-systems concerned does not require a single simulation need, but a set of different needs. This has a direct consequence on the nature of the associated simulation models: these models are designed with different required granularity and accuracy levels.

For the same component, it is possible to consider several modeling levels, from the simplest to the most complex. The simplest modeling for a *"sensor"* system is to represent its expected role from the viewpoint of the entire system. In other words, this *"simple"* level of modeling, also called *"systemic"*, represents the direct capture of the state of the environment this sensor is expected to produce, regardless of its actual performance. For example, a *"systemic"* model of the RADAR sensor can potentially return the exact position and speed of each individual object in a given area (sensor perception area). We sometimes speak of an *"idealized"* sensor model. This type of model can be deployed to verify the underlying logic of situation analysis and decision-making, for example.

On the other hand, it is accepted that this level of *"idealized"* modeling does not address the simulation of function performance in real conditions. Since the performance of sensors is intrinsically limited by their physical domain, their operating technology and their environmental conditions, these parameters must be taken into account to simulate the real performance of the Connected Automated Vehicle's performance. In order to do this, it is necessary to carry out an analysis of the sensor in question, so as to determine and quantify its sensitivity to these parameters, and to infer the most appropriate models for them.

4.4.3. *Optical sensors*

4.4.3.1. *Working principle*

The optical camera is an electromagnetic signal sensor relying on light captures from the environment, in order to form images. The part of the electromagnetic spectrum which is commonly used corresponds to the sensitive silicon area (visible and near infrared, for example, from 400 to around 1300 nm). This use of silicon makes it possible to obtain sensors with a good sensitivity at relatively low costs. Most cameras used in the automotive industry are said to be *"passive"*, in the sense that they simply capture the light produced by the environment. Some other cameras, called *"active"* cameras, illuminate the environment within a certain frequency range, and collect the resulting light or energy.

The camera uses the physical principles underlying human vision, which makes the data from this sensor easy to interpret by humans. Nevertheless, interpretation and understanding tasks are relatively difficult for image processing algorithms, since they have to detect and, above all, recognize objects. In addition, this sensor is highly sensitive to the illumination conditions resulting from natural lighting (meteorological conditions, in particular), artificial lighting (public lighting and vehicle lighting), and obviously, the optical characteristics of objects.

The sensitivity of optical sensors to the "*visible*" spectrum generates a particular sensitivity to climatic conditions. Unlike other sensor technologies (RADAR, LiDAR), the detection performance of an optical sensor is usually very different, depending on the perception configuration: visibility distance, lighting level, precipitation level, changes in the reflectivity of surfaces. Embedded applications for detecting road markings or obstacles can be strongly disturbed by these particular perception conditions. As a consequence, it seems important to be able to simulate these "*degrading*" and "*adverse*" conditions in the sensor model in order to be as close as possible to reality.

The proposed optical sensor model breaks down processing into two distinct parts: the optical system and the optical sensor. The first part concerns the initial rendering stage, which makes it possible to obtain one or more images in ideal conditions. This can be likened to the so-called projection-perspective "*pinhole*". The second stage simulates general flaws related to sensor components.

4.4.3.2. Image rendering simulation

This first calculation stage is carried out by relying heavily on the capacities of the graphics engine of the simulation platform. A dedicated lighting calculation module takes into account the lighting distribution of light sources, as well as the physical surface condition of the materials of the objects which make up the scene. The apparent luminance of the objects is thus evaluated, as well as the angular luminance function at the camera's input. In automotive applications and road scenes, the types of materials used are generally plastic, metal, rough surfaces such as bitumen, and backscattering materials (for markings and traffic signs). In order to cover the properties of all these types of materials, models by Lafortune (Lafortune 1996, 1997), Cook-Torrance (Cook-Torrance 1982), and Blinn-Phong (Blinn-Phong 1976; Blinn 1977, 1982) were implanted using OpenGL *shaders*. The interesting contribution of these *shaders* is the ability to perform lighting calculations for each camera pixel (raster rendering), instead of calculating each geometric vertex in the scene. It is therefore possible to simulate reflections whose apparent width is much smaller than the characteristic dimension of the objects' mesh. This makes it possible to significantly optimize the computational cost associated with such objects.

Figure 4.16. *Illustrations of effects related to lighting in pixel rendering, such as the fine simulation of night headlights lighting (left), and the simulation of a situation in low sunlight conditions (right)*

4.4.3.3. Modeling of effects related to sensor components

The second stage for modeling optical sensors simulates general flaws in the sensor's components, such as distortion, chromaticism, and losses of light intensity. An extra stage is also integrated to take into account more particular flaws such as lens flare, glare, etc. These mechanisms and filters correspond to the different functions and modules which are intrinsic to the camera (Figure 4.17): image capturing process (shutter mechanism, motion blur), spectral efficiency, sensor response, light flow, amplification stage, gain and exposure control, and finally data conversion from analog to digital mode.

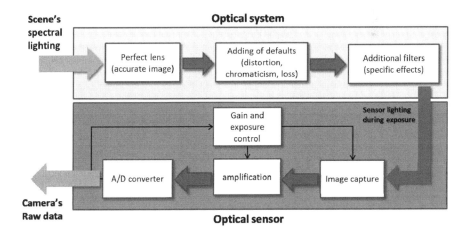

Figure 4.17. *Schematic diagram of the optical camera model*

To calibrate the camera model, that is to say, to obtain all the parameters corresponding to a particular camera model, two approaches are possible:

– obtaining the parameters from the interpretation of the camera's technical data sheet, or the camera's subcomponents;

– a calibration procedure on a test bench, when the camera's characteristics are unknown *a priori*.

It was thus possible to compare data from a set of real cameras with simulated cameras. Nevertheless, the procedure implemented required an identical reproduction of the camera's environment (which can be either an open environment or the test bench). The real and simulated calibration targets used for testing some of these cameras are shown in Figure 4.18. The errors obtained between the simulated cameras and the real cameras are negligible. For more information, see (Gruyer 2012; Grapinet 2012).

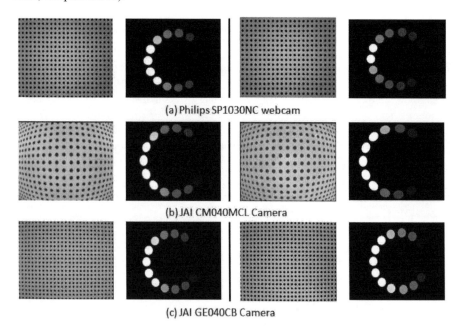

(a) Philips SP1030NC webcam

(b) JAI CM040MCL Camera

(c) JAI GE040CB Camera

Figure 4.18. *Simulation and real camera data (calibration targets) (Gruyer 2012)*

4.4.4. *LIght Detection And Ranging (LIDAR)*

4.4.4.1. *Working principle*

A LIDAR sensor (acronym for *"LIght Detection And Ranging"*) is, by definition, an active sensor: it illuminates the scene usually employing laser beams, and collects the energy sent back by the scene. Under ideal conditions of use, the beam is reflected directly from the objects (a single bounce), and the time elapsed between the emission of the beam and the signal return provides a highly accurate measure of the distance to these objects. Angular scanning and the use of multiple beams (multi-layer sensors) make it possible to obtain measurements for an extensive angular range (horizontal and vertical).

Usually operating in the near infrared, this type of sensor is highly sensitive to the reflectivity of objects (reflection coefficient related to the angular configuration, granularity, and color of the material) and to the configuration of the propagation channel (climatic conditions (rain, fog, etc.), dust and particles in suspension), which can alter their performance (range, energy), or even make them inoperative. Simulating the performance level of a LIDAR depending on driving conditions makes it possible to predict the performance of the detection chain using LIDAR sensors, in a much more accurate way.

LIDARs are known to be expensive sensors, which limits their current deployment. Nevertheless, various technological innovations have made it possible to reduce their cost, to build more compact sensors, more robust to vibrations and failures, as is the case of so-called *Solid State* LIDARs compared to mechanical scanning LIDAR sensors.

4.4.4.2. *Simulation model*

LIDAR modeling takes into account both the intrinsic and extrinsic aspects of the sensor. For the intrinsic aspect, the acquisition mechanism and the optical signal processing part are modeled. As regards the extrinsic aspect, the propagation channel and the light propagation conditions in the environment of the sensor are considered in the equation.

The sensor is modeled so as to be able to represent both scanning and *Solid State* LIDAR sensors. The distribution of the emission and acquisition directions of the laser beams is adjustable, as is the energy of these beams. The model can be adapted for high resolution LIDARs with a restricted field, or for 360° LIDARs.

Figure 4.19. *Simulation of a Velodyne LIDAR sensor model in Pro-SiVIC and representation of the points cloud in clear weather*

The distribution of the LIDAR beam directions provides a representation of the horizontal and vertical resolution of the sensor (Figure 4.19). Several levels of simulations have been implemented in the physical platform, some of which use the availability of graphics power to speed up the computation time. While the simple models are content to simulate only the geometric aspects of detection with LIDAR, the more detailed ones take into account the propagation conditions of LASER beams and the type of targets the LIDAR perceives.

Figure 4.20. *Simulation of a Velodyne LIDAR sensor model in Pro-SiVIC and representation of the points cloud in moderate rainy weather*

In addition to the geometric aspects of the beam emission, the model takes into account the power emitted by the LIDAR, the atmospheric conditions, and the physical parameters inherently specific to the materials illuminated by LASER beams (Figure 4.20). In this way, the proposed model makes it possible to faithfully simulate the levels of detection performance associated with a LIDAR sensor, depending on its detection capacities, even in degrading meteorological conditions (rain, fog, dust, particles).

4.4.5. *RAdio Detection And Ranging (RADAR)*

4.4.5.1. *Working principle*

RADARs for "*automotive*" applications are sensors which emit electromagnetic waves with a very high frequency carrier (24 Ghz, 77 and 79 Ghz) and a modulated signal (FSK, FMCW, etc.) to enable the processing of the energy retrieved back to a receiving antenna, and thus estimate object distance and speed. Thanks to their nature and technology, RADAR sensors are sensitive to certain parameters and certain electromagnetic disturbers in the environment. Unlike cameras and LIDARs, this type of sensor is not affected by climatic disturbances and lighting conditions in the environment. Generally, the functional modeling of these sensors can be broken down into two main parts. The first one characterizes the physical sensor (antenna, equipment, electronics, signal processing) and the technology used, whereas the second one characterizes the electromagnetic propagation of the waves emitted by the transmitting antenna, and their interactions with the environment before reaching the receiving antenna.

4.4.5.2. *Simulation model*

Stricto sensu, the "*physical*" sensor model can be broken down into a set of functional blocks characterizing the different modules integrated in the RADAR sensor for generating and managing the signal (Figure 4.21). This generic model can be specialized for the automotive domain (and other sectors) thanks to the use of VCO components (*Voltage Controlled Oscillator*), which generate the modulated signal, and the COMP (*Computer*) component, which processes the signal for extracting useful environment-related data. Several technologies, called academic technologies (FSK – FMCW), are available for the modeling of an "*automotive*" sensor. This model is sufficiently generic to integrate new generations of sensors.

The second part of the model focuses on the propagation of electromagnetic waves in the environment. Four remarkable characteristics of the wave trajectories are taken into account in our model:

– the power of the received signal, or ratio between the received/transmitted signal. This information defines the level of energy sent back to the antenna after interacting with the environment;

– the Doppler effect which characterizes the frequency shift of the signal, due to the relative speed difference among the scene objects having interacted with the signal;

– the phase shift, due to the distance traveled;

– the polarization of the wave which determines whether or not the path is taken into account when returning to the antenna.

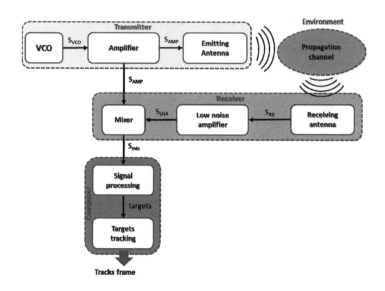

Figure 4.21. *Functional diagram of a RADAR*

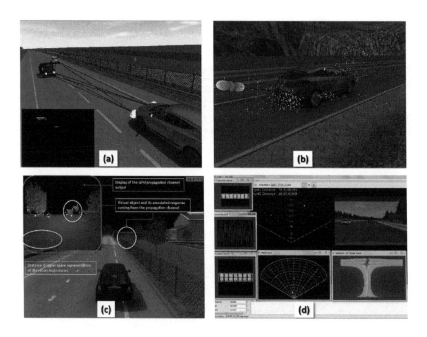

Figure 4.22. *Level 1 RADAR modeling: (a) SER in the form of a parametric lobe, (b) SER in the form of a bi-static data base, (c) scene lighting thanks to physico-realistic modeling, (d) antenna and signal processing*

Four levels of modeling have been proposed, using varying complexity representations for the propagation of an electromagnetic wave in a road environment.

In particular, a modeling level makes it possible to combine 3D electromagnetic calculation methods with a real-time simulation of the radar system. In that way, it is possible to increase the representativeness of the simulation, by taking into account the fine characteristics of antenna radiation, the integration of the RADAR antenna in the vehicle and the target's RADAR response.

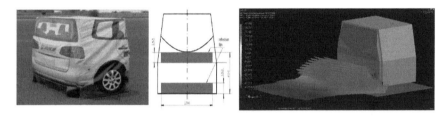

Figure 4.23. *Illustration of a Euro-NCAP 2016 target (left), and visualization of the simulation of its electromagnetic response (right), simulated by CEM One*

The characteristic response of objects as an angular distribution of SER (*Surface Equivalent Radar*) can be generated and used with dedicated electromagnetic calculation tools, such as *CEM One* (software marketed by ESI Group, Figure 4.23). The generation of these SERs is accomplished by using asymptotic high frequency computational methods. The main advantage of this type of approach is to unify the purely physical simulation results, which require significant computation time and a substantial hardware infrastructure, with virtually real-time simulation on individual workstations (Kedzia 2016; Hadj-Bachir 2019b).

4.4.6. *Global Navigation Satellite System (GNSS)*

For the development of ADAS and, more recently, automated vehicles, the fact of knowing the position and dynamic state of the ego-vehicle is essential. To achieve this function, getting to know the absolute positioning of a vehicle, proprioceptive sensors are generally used, comprising odometers, inertial units (INS and IMU), and, most importantly, a GNSS.

4.4.6.1. *Working principle*

To position oneself using a GPS system, in theory, it is enough to know the position of the visible satellites by a receiver positioned on the planet's surface, as well as the distance measures (called pseudo-distances) between these satellites and the receiver. Then a simple triangulation makes it possible to calculate the receiver's position. As many sources of errors affect this information, errors are introduced in the position's estimate (errors may range from a few meters to a few tens of meters). To obtain a realistic simulation, it is therefore necessary to simulate this information, as well as the error models disturbing the calculation of the pseudo-distances.

Pseudo-distances measure the distances between visible satellites and the receiver. They are deduced from the time of flight it takes for the signals transmitted by the satellites to reach the receiver. These signals go through the different atmosphere layers (including the ionosphere and the troposphere), and can be reflected by buildings (*"urban canyon"* problem), before reaching the receiver. Then the receiver evaluates their time of flight, and finally, by applying the speed of light, it is possible to calculate the resulting pseudo-distances. These times of flight are biased because the clocks of the receiver and of the transmitters (satellites) are not synchronous, and the traveling time of waves is overestimated, since their propagations are disturbed and delayed by the propagation channel (atmosphere and reflections). It is generally agreed that measurement errors on pseudo-distances are produced by four main sources: errors introduced by crossing the ionosphere and the troposphere, multi-paths and clock offsets.

In order to be able to evaluate and correct the ionospheric and tropospheric errors, a set of correction models is available. Among these models, we can quote the *Klobuchar, Hopfield, Marini, Saastamoin, Cent, Goad* and *Goodman* models. Some of these models are embedded in GPS receivers in order to correct atmospheric errors in real time. The calculation of the pseudo-distances is obtained using the modules and functions listed in Figure 4.24.

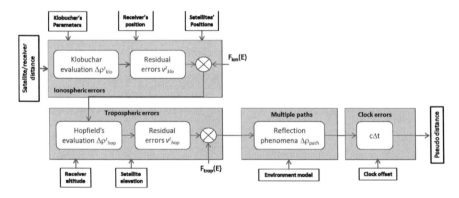

Figure 4.24. *Stages for the calculation and simulation of a pseudo distance*

4.4.6.2. *Simulation model*

The simulation of the satellite constellation requires calculating the position of all the satellites for a given date. Thanks to the ephemeris files available from the International GNSS Service website (https://kb.igs.org/hc/en-us/articles/115003935351), it is possible to calculate the position of the satellites either by performing a Lagrangian interpolation or by using the orbital parameters deduced from the ephemeris. In practice, satellite trajectories (and therefore their positions) are tainted by errors. In the simulator integrated into Pro-SiVIC, the only trajectory errors taken into account will only depend on the accuracy of the ephemeris.

If we express these modelings in an equation, the calculation of a pseudo distance is obtained using the following equation:

$$\rho_m = \rho + c\Delta t + F_{ion}(E)(\Delta\rho^2_{klo} + v^2_{klo}) + F_{tro}(E)(\Delta\rho^2_{hop} + v^2_{hop}) + \Delta\rho_{path}$$

where:

1) ρ_m is the pseudo distance measured by the receiver;

2) ρ is the distance separating the satellite and the receiver;

3) Δt is the clock offset between the satellite and the receiver;

4) E is the satellite's elevation;

5) $F_{ion}(E)\Delta\rho^2_{ion} = F_{ion}(E)(\Delta\rho^2_{klo} + v^2_{klo})$ is the ionospheric error;

6) $F_{tro}(E)\Delta\rho^2_{tro} = F_{tro}(E)(\Delta\rho^2_{hop} + v^2_{hop})$ is the tropospheric error;

7) $\Delta\rho_{path}$ represents the errors introduced by multi-paths.

The simulation of a GPS receiver can then be broken down into three stages:

– estimate the position inferred from the pseudo-distances;

– calculate the parameters describing the accuracy and consistency of the estimate;

– format the result obtained in one of the NMEA frame format (GGA, GSV, etc.).

In the current modeling, the errors related to the electronics of the receiver were not taken into account.

The integration of the GPS model mainly relies on two components, and on the use of a GPS simulation library designed during the EMOTIVE project (*Environment MOdeling for percepTive Intelligent VEhicles*) by LASMEA (eMotive 2010). The first component is dedicated to the actual GPS simulation, whereas the second component deals with the display of the results produced by the GPS simulation.

Figure 4.25 shows a screenshot of the display of GPS data in Pro-SiVIC. This visualization includes a *"compass"* whose needle is the vehicle equipped with GPS receiver, cardinal points, a *"scale"*, the calculated position (*"white cross"*), as well as three NMEA frames.

In addition, in order to consider occultations and the problems of multiple reflections, a render layer uses *"bounding boxes"* to model vertical objects. In order to build this modeling of the environment in the form of ellipsoid oriented *"bounding boxes"*, a Principal Component Analysis was adopted.

The following figure shows a use case on the Versailles-Satory test tracks. This example shows the results obtained with a simulated GPS compared with a real RTK GPS and a real natural Trimble GPS. By projecting data from the Trimble GPS and the simulated GPS into the virtual environment, it is clear that the results obtained are comparable.

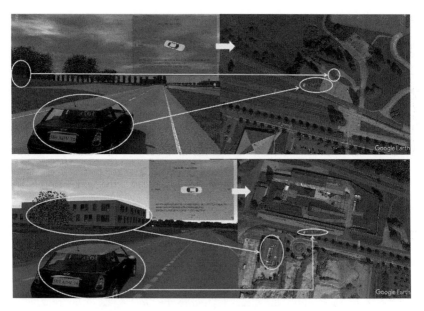

Figure 4.25. *Position display modes in Pro-SiVIC and projection in a Google Earth top view*

Figure 4.26. *Simulation of GPS data in Pro-SiVIC at
the same time as real GPS data (RTK and TRIMBLE)*

4.5. Connectivity and means of communication

4.5.1. *State of the art*

In recent years, more and more work has been carried out on the development of cooperative systems. For this, various communication media are used. The best known is probably WiFi (*Wireless Fidelity*) for VANet (*Vehicular Ad-Hoc Network*). In order to be able to prototype these cooperative and communicating applications, it was imperative to be able to model these media in the Pro-SiVIC platform. This type of modeling began with the development of short-range *transponder*-type communicating beacons. Even though the data frames used were the same as those produced by a real transponder, the propagation channel was greatly simplified, using only a range limited omnidirectional transmission model. No energy mitigation, message loss, or transmission delay mechanism was modeled. In addition, no masking mechanism was available. However, this model was sufficient for simulating a very short-range transponder usable for speed regulation, or for the production of punctual information about infrastructure constraints and events (speed limit, alert, road works etc.).

However, since the European CVIS and SafeSpot projects, it has become unquestionable that the scientific community working on driving assistance systems would increasingly need to resort to dedicated communication network simulators (with advanced functionalities and validated models), in order to manage inter-vehicle (V2V) or vehicle-infrastructure (V2I) communications. It thus became necessary to develop functionalities in Pro-SiVIC for simulating means of communication and for tackling development/study problems about the impact of automated and connected vehicles. With its interoperability capabilities, the development of Pro-SiVIC has naturally turned towards its coupling with already existing third-party simulators.

There were already other works around the coupling of a network simulator with a road traffic simulator, such as those carried out within the framework of the iTETRIS initiative (http://ict-itetris.eu/) (Rondinone 2013). However, we noticed that the propagation models implemented by these solutions were statistical models which barely made it possible to model the phenomena observed in real conditions (such as masking). These solutions were better suited for the study of macroscopic phenomena (routing problem in a VANet), and less suited for ego-centered microscopic problems. It therefore became necessary to find another solution to approach the study of cooperative systems in which an ego-vehicle must alert its neighbors about a dangerous event that could lead to emergency braking.

Ever since the development of FUI EMOTIVE, ANR PRCI CooPerCom, and FUI SINETIC projects, the Pro-SiVIC platform has included modules for simulating an electromagnetic propagation channel based on detailed data from the road scene. Until now, these modules have been used for the simulation of RADAR type sensors. These modules have been adapted and modified to become more generic and applicable for the simulation of wireless communications, such as WiFi (802.11p).

As for RADAR, several levels of modeling have been proposed to better adapt to the level of complexity required for simulating means of communication. Except for transponders, two levels of 802.11p type telecommunications modeling have been implemented in the Pro-SiVIC platform. In addition, this platform dedicated to cooperative and communicating systems, was called SiVIC-MobiCoop (Gruyer 2013, 2018).

4.5.2. *Statistical model of the propagation channel*

The first level of modeling integrates the statistical models developed within the doctoral research works of S. Demmel (2014). This modeling has been obtained from the statistical analysis of a large amount of inter-vehicle communication data (using 802.11p modems) with the aim of estimating the Frame Loss Rate transmitted as a function of the emission distance and relative speed between transmitter and receiver. In this model, the communication range indicator can be interpreted as a subset of frame loss indicators. In fact, the maximum range can quite simply express the distance at which frame loss is total. We focused on the modeling of this indicator from the experimental data obtained on the high speed track in the Satory test tracks. Simply put, the input variables of our model are the distance and the relative speed between a transmitter and a receiver (both on a vehicle and in a roadside unit). The output of our model is the probability of experiencing frame loss. This model is capable of reproducing all of our experimental data, but also makes it possible to generate "new" (unmeasured) data, which will be plausible given the observed measures. Due to the environment used for our measurement campaign, our model is suitable for open motorway environments, as well as for rural roads. However, this model is not suitable to operate in urban settings where we have LoS (*Line of Sight*) conditions which are not guaranteed, as well as a high density of objects causing interference and multi-reflections. This model (Demmel 2014) is defined in a generic and parametric form which integrates the pooling of several sub-models, a part of which corresponds to the model taking into account the interference caused by the reflection of radio waves on the ground. This model assumes that no countermeasure is employed to mitigate the frame loss phenomenon due to interference.

4.5.3. *Multi-platform physico-realistic model*

The second level of modeling (Gruyer 2018; Ben Jemaa 2016), which is much more complex and realistic, does not require knowledge about the transmission capacities in the environment, *a priori*. It is based on the use of 3 sets of modules and functionalities: firstly, the NS3 library dedicated to the simulation of communication networks (the 7 layers of the OSI model, the routing mechanisms, mobility graphs, etc.), secondly, the adaptation of the propagation channel defined in the Pro-SiVIC platform for electromagnetic sensors of RADAR types, and thirdly, the DDS communication bus which ensures the interoperability of applications and is used for interconnecting Pro-SiVIC with third-party applications involved in this simulation. The interest of this approach is twofold, since it makes it possible to overcome the knowledge about the reception capabilities in an environment, while making it possible to establish a complete, generic, and interoperable telecommunications simulation platform, comprising all the technologies developed by the scientific community (partly available in the NS3 library).

More specifically, the propagation channel integrated in Pro-SiVIC makes it possible to calculate the characteristics of the main wave paths between two antennas at any given moment, as well as mobility data (gain and attenuation of the energy on the path, path length, Doppler effect, position, orientation and kinematic torsor of the visited points) (Figure 4.28). All the software modules involved in the simulation of communications (NS-3, Pro-SiVIC propagation channel, communication bus, frame processing platform, etc.) are presented in Figure 4.27.

In addition, with these models, it is now possible to set up prototyping scenarios of cooperative applications no longer made up of a single vehicle equipped with means of perception, but of a set of vehicles which can perceive the environment and interact with other mobiles. Potentially, and with only a few modifications, this platform allows us to mix up messages from real vehicles and virtual vehicles evolving in the Pro-SiVIC platform, within the same communication flow. This feature can be very interesting for simulating potentially dangerous situations without endangering real material and human resources.

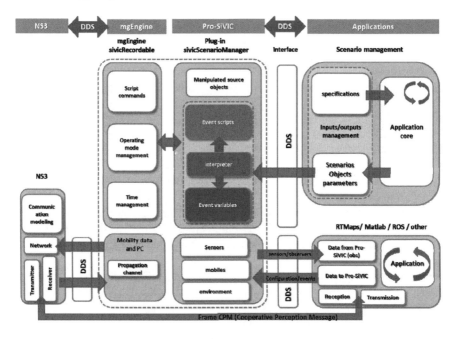

Figure 4.27. *Functional diagram of Pro-SiVIC-MobiCoop in real-time operation, with the interconnections to NS-3 and other third-party applications*

The frames transmitted are based on the description given in the document produced by HERE in 2015 (Thandavarayan *et. al* 2020).

Figure 4.28. *V2V communication in Pro-SiVIC with omnidirectional antennas and adapted propagation channel*

4.6. Some relevant use cases

4.6.1. *Graphic resources*

So far, this chapter has presented the main functionalities required for the prototyping, testing, evaluating and validating ADAS, as well as applications related to the design of automated and connected mobility services. In order to remain as close as possible to reality, a set of graphic resources was created, which can reproduce road scenes in general, and the Versailles-Satory test tracks, in particular. Figures 4.29 and 4.30 present the type of physico-realistic modeling of the environment obtained. The applications that will be presented in this section were tested in this environment.

Figure 4.29. *Level of detail in the renderings of the Versailles-Satory "Main Road" and "Speed" tracks*

4.6.2. *Communication and overall risk*

In order to implement communicating-systems applications, it is important to have a faithful model of the behavior of the transmission of a data frame between two antennas: the transmitting and the receiving one. In fact, it is necessary to know how to model the propagation channel. Numerous studies on the simulation and use of the 802.11p standard for VANET-type communications rely on NS2 or NS3-type modeling libraries. It is known that the complexity of the road environment implies that in some cases the performance measurements of 802.11p are likely to diverge considerably from theoretical models. Even if the simulators relying on the NS3 library can be configured to use an indirect propagation model acknowledging ground reflection, the experimental results conducted in Sébastien Demmel's doctoral thesis suggest that this model is not always suitable for representing certain variations in performance, which we, in turn, were able to measure and observe. It is for this reason that an empirical and statistical modeling has been brought forward. This is a good way to supplement and improve the results of simulations. It is this modeling which was used in this application.

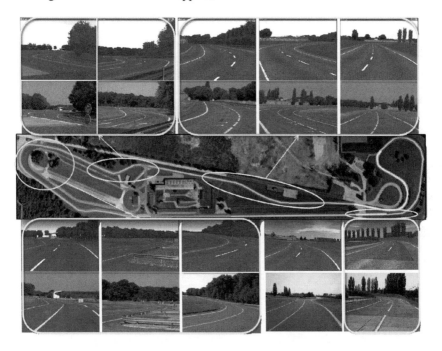

Figure 4.30. *Level of detail in the renderings of the Versailles-Satory "Road" track*

The implementation of this model in the Pro-SiVIC platform made it possible to reproduce the "*emergency stop*" scenario in a line of vehicles in a much more realistic way (Gruyer 2013). The purpose of this scenario is to quantitatively show the interest and benefit of deploying means of communication to reduce the risks of collision, and their severity. This study requires having an experimental plan where the equipment rate (1 to all vehicles), as well as the spatial distribution of this equipment may vary. In this simulation, vehicles and their dynamics are reproduced in a very realistic way. In addition, communications incorporate the imperfections of the 802.11p model (ground reflection, frame loss rate as a function of distance and relative speed, transmission latency, etc.). Finally, each vehicle has a reaction time (system and human, depending on the case) which can be variable, and the ability to detect frontal obstacles. In this scenario, the vehicles also perform emergency braking, based on their braking capacity and tire grip coefficient. This makes it possible to obtain behaviors which closely resemble reality. The functional diagram of this complex application is shown in Figure 4.31.

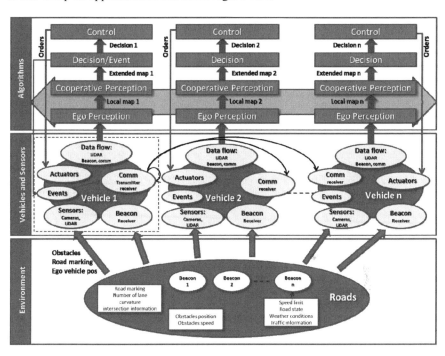

Figure 4.31. *Functional diagram of the communication/ global risk estimation application*

After multiple tests with variable equipment rates, and depending on this equipment rate, a very significant reduction in the number of collisions was observed and measured. However, even if the reduction in the number of collisions was pertinently observed and quantified, we also noticed that the EES (*Equivalent Energy Speed*) remained almost constant (Figure 4.33). This means that the fewer collisions there were, the higher the severity of the remaining collisions could be in some configurations. This result deserves further study because it raises serious questions about the strategies to be implemented when collisions occur. However, so far, these tests have only been carried out for alert transmission-type scenarios (Figure 4.32).

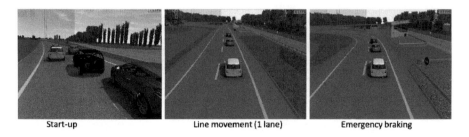

| Start-up | Line movement (1 lane) | Emergency braking |

Figure 4.32. *The three stages of the scenario: starting on two lanes, converging towards a single lane queue, and finally emergency braking of the "leader" vehicle*

Thanks to these studies, it is clear that the use of communication means produces a significant impact on the reduction of collisions (Gruyer 2013). This result was an important first step to validate the development of cooperative systems. In the following stage, research was extended by seeking a solution not only to limit the number of collisions, but also for anticipating risky situations. The question is whether we can guarantee a maximum safety level or a minimum risk level. In this case, sending an alert message is no longer sufficient. It is necessary to be able to handle more complex information about the current situation. This information corresponds to the whole of, or to part of the data coming from the ego-vehicle's Local Perception Dynamic Map.

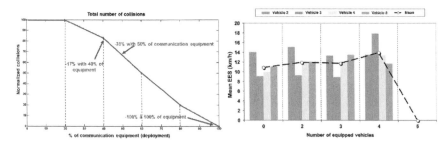

Figure 4.33. *Normalized number of collisions for the complete line and per vehicle, depending on the equipment rate, and average EES for each vehicle and per equipment rate*

Each vehicle equipped with embedded perception and means of communication will transmit its local knowledge to other vehicles equipped with means of communication. When a vehicle or an infrastructure manager takes into account various local perception maps, it is then possible to build extended perception maps, to gain awareness about the events and the existence of the road scene's key components (obstacles, road, ego-vehicle, environment, driver). This extended perception can also manifest as a greater perception range, as well as an enriched description of the road scene's key components. The increase in this perception range will be very important for anticipating and predicting events which could lead to accident-causing situations. The interpretation of these future situations will allow decisions which could lead to a modification of the ego-vehicle's behavior, in order to minimize the level of risk. In this new study, the goal is mainly to find out whether it is possible, and particularly relevant, to estimate an overall level of risk from this extended perception. In addition, it seemed logical to quantify the contribution, if any, of this global risk compared to the local risk. To carry out this study, we had to choose or build a risk criterion. Our choice fell on the use of a classic risk metric based on *Time To Collision* (TTC). The TTC represents the quantization of the remaining time interval at a certain moment t before a future collision between two objects (vehicles or others), unless they modify their current behavior, depending on inter-distance, speed, acceleration and/or direction. In fact, TTC is a magnitude often replacing the notion of risk. In order to compare the performance of the local risk estimate to the global risk based on the use of extended perception, it seemed more relevant and more appropriate to simply create one single risk value, describing the dangerousness of the situation, instead of a set of risks calculated for each vehicle. For each test over time, we implemented two risk values. First, the overall risk $R_{g,\,x}$ calculated for a single vehicle x. Then we will have the overall perceived risk R_g for all vehicles. The higher R_g, the more dangerous the driving situation. In the scenarios implemented, the first four vehicles transmit their positioning at 2 Hz, with a maximum latency of 5 ms. The ego-vehicle is at the end of the line and updates its own localization and its local perception map at a frequency of 10 Hz and 20 Hz, respectively. The extended perception map itself, is updated at a frequency of 10 Hz. The local ego-vehicle perception map is obtained by applying target detection and tracking, with LiDAR technology. Its localization uses a nonlinear Kalman filter (*Extended Kalman Filter*).

If we set a critical danger threshold at the risk value 0.7, then the driver will be alerted 5 seconds before the collision. This warning time is short and will only enable the driver to react (emergency braking or avoidance maneuver), but not to anticipate the situation, in order to implement a maneuver which minimizes the risk. This anticipation factor is important to limit the occurrence of a possible collision with a rear vehicle. If we now observe the results presented in Figure 4.34, and obtained with the use of cooperative applications, we can observe glaring event modifications. Indeed, halfway through the scenario, the local risk between vehicles 4 and 5 is close to the dangerousness threshold, whereas the overall risk remains below this threshold. In this situation, vehicle 5 actually considers itself in a very risky situation, whereas this is not the case for the other vehicles. This therefore attenuates the importance of local risk perception, in relation to a much more global perception. The two risk indicators (local and global) are therefore in contradiction, but nonetheless perfectly reflect the situation from two different viewpoints. At the moment when the first vehicle performs emergency braking, we can observe that the risk level of vehicle 2 increases almost immediately. Then, the other vehicles also see their respective risk increase in cascade, which translates into a series of emergency-braking reactions. However, by observing the crossing of the local and global risk indicators above the dangerousness threshold, one can see that the global risk crosses this threshold approximately 7 seconds before the local risk. This remark is very important because it informs us that vehicle 5 no longer has only 7 seconds to react, but twice as long, that is to say, 14 seconds. This gives us sufficient time to make better decisions and initiate maneuvers minimizing global and local risks. This behavior is important in the case of the disengagement of a driving automated system requiring the driver to take back control. The longer the driver has time to perceive, interpret and understand the situation, the more he will be able to act in a suitable and safe manner.

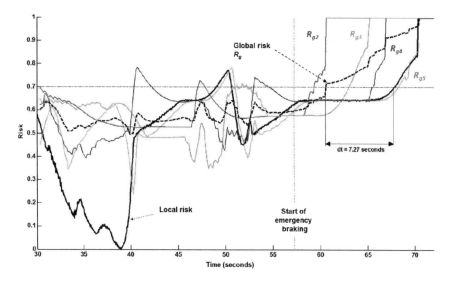

Figure 4.34. *Local risk and global risk in the vehicle line scenario*

4.6.3. *Automated parking maneuver*

This application was developed as part of Sungwoo Choi's doctoral thesis at Mines ParisTech (Choi 2010), the purpose of which was to propose an automated parking maneuver algorithm, which would be optimal in terms of the number of maneuvers to be performed (Choi 2011). This application was first prototyped in Matlab (Figure 4.35), and was later implemented in RTMaps with Pro-SiVIC, for the modeling and simulation of vehicles, sensors, and actuators.

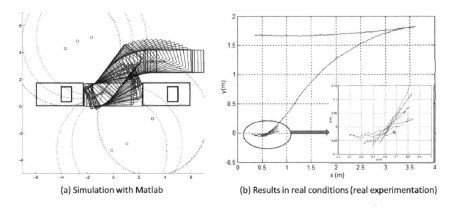

(a) Simulation with Matlab   (b) Results in real conditions (real experimentation)

Figure 4.35. *Implementation of the automated and optimal parking application in Matlab and its actual results (Choi 2011)*

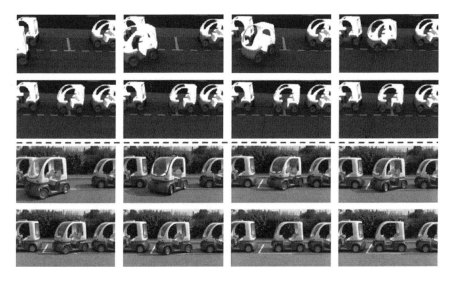

Figure 4.36. *Implementation of the automated and optimal parking application with Pro-SiVIC/RTMaps and in real conditions (Gruyer 2014a)*

Figure 4.36 shows screenshots of the experiments performed within the Pro-SiVIC environment, with real application modules. Afterwards, these modules were implemented on INRIA's Cycabs, in Rocquencourt. During these tests and the development cycle, ranging from the simulation phase to real life conditions, we were able to notice the same behaviors during simulation and in real life experiments. This enabled us to validate the concept, as well as the representative character, of simulation for evaluating this type of application. This automated parking concept has been generalized for parallel, perpendicular, and herringbone parking maneuvers. Figures 4.37 and 4.38 present these simulated experiments with a complex model of vehicle dynamics and with embedded telemetry sensors for detecting a sufficiently large free space.

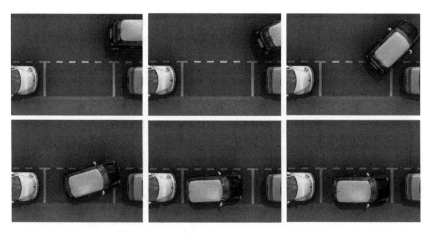

Figure 4.37. *Automated parallel parking maneuver*

Figure 4.38. *Automated perpendicular parking maneuver*

4.6.4. *Co-pilot and automated driving*

In the previous application, control was only applied to a single vehicle and in a local low speed configuration. As part of the European FP7-ICT HAVE-it project (*Highly Automated Vehicles for Intelligent Transport*) and the thesis work of Benoit Vanholme (2012), this research scope was extended with the goal of controlling a fleet of vehicles (longitudinal and lateral maneuvers) to study the deployment of highly automated vehicles at high speed in a motorway environment. In this work (Figure 4.39), a virtual co-pilot model was proposed and developed taking into account the information from the extended dynamic perception map (near and far markings and obstacles), as well as a set of rules (human, system, traffic). For example, traffic rules correspond to the Highway Code rules (speed limits, respect for road markings, mandatory overtaking on the left, etc.). The use of these data and constraints has made it possible to propose longitudinal and lateral driving instructions, minimizing a set of criteria including the risk and cost of a situation. This work has been presented in detail in (Vanholme 2013). In this application, apart from having to operate at high speed in a motorway environment, the co-pilots had to be able to implement several types of driving behavior (sporting, comfortable, normal).

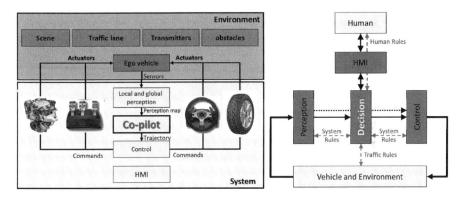

Figure 4.39. *Modules and rules acknowledged by the design of a virtual co-pilot*

The advantage of using simulation to prototype this type of central virtual co-piloting function for automated vehicles is to be able to concentrate efforts on only part of the development (a single function) and to leave the rest of the functions in "perfect" mode. In our case, we focused our efforts on the development of the co-pilot itself, and we used a "*perfect*" environment perception. For the perception task, we actually used the "*reference*" sensors (road markings and obstacles), and not the processing stages derived from a set of simulated realistic sensors. Thus, we did not have to question the quality of the perception and we could concentrate, initially, only on the development of the co-pilot. The functional architecture developed for this platform is presented in Figures 4.39 and 4.40. In this second figure, we can see that the application has been designed with three levels.

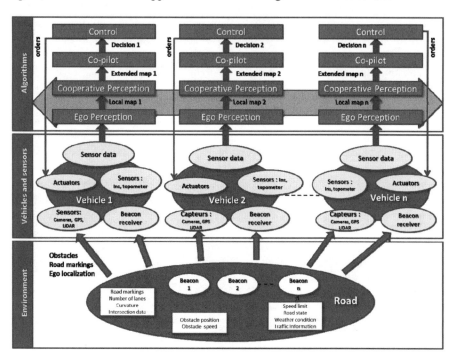

Figure 4.40. *Functional architecture of the co-pilot application (HAVE-it project)*

The first level concerns the environment and roadside equipment, such as beacons transmitting speed limits (Gruyer 2009). The second level concerns the vehicle with its actuators, as well as the perception (sensors) and communication equipment. Finally, the third level is dedicated to the different applications used for enabling the automation of high speed driving. In this "*application*" layer, four processing levels are needed: perception of the local and close environment, cooperative perception or extended perception, the virtual co-pilot, and the control making it possible to generate the commands/orders to act on the vehicle's actuators. For this application, demonstrations involving nine automated vehicles were implemented. Each vehicle had its own co-pilot configured for a specific driving mode (sporting, normal, comfortable). In addition, a driving sharing feature was integrated, enabling the driver to take back control of the driving task for one of the vehicles at any moment. This mechanism made it possible to observe the impact of human behavior in an area where only autonomous vehicles circulate. The robustness of this demonstration was also appreciated by making this fleet of vehicles operate for long hours, without any interruption (more than 400 km traveled by the vehicle, in normal mode). In these long experiments, it was possible to observe that different potentially opposing driving modes (comfortable and sporting) could coexist without provoking a collision or "near accident" situations (event and high risk situation likely to create an accident).

Figure 4.41. *Use of the co-pilot developed in the European Have-It project on the Satory tracks, in a real vehicle and on the Pro-SiVIC platform*

Later on, the co-pilot was tested in real conditions on a dedicated hardware architecture and on one of the LIVIC prototypes (Figure 4.41). This very interesting study made it possible to validate the use of a simulation platform as a means for prototyping and validating all the design stages of a complex application devoted to maneuver/trajectory planning and to the high speed automated driving systems. Other applications were subsequently developed and tested with the same operating mode for vehicle control problems via *"model-free"* approaches (Menhour 2018), for the management of cooperative platoons and for shuttle management in dedicated environments.

4.6.5. *Eco-mobility and eco-responsible driving profile*

So far, we have presented Pro-SiVIC as a platform mainly used for sensor modeling and for active ADAS prototyping. With the marketing of immersive virtual reality helmets (OCULUS, HTC Vive), it seemed relevant to interface Pro-SiVIC with this new technology. The goal was to extend the functionality domain of Pro-SiVIC and obtain a full 3D immersive application for a real driver involved in the simulation loop (Figure 4.42). In addition, thanks to the works of Olivier Orfila (2015, 2017b) on the issue of eco-driving and eco-mobility, it was possible to implement a new sensor for measuring the energy consumption of a vehicle, in Pro-SiVIC. With this new sensor and the use of a virtual reality helmet, it was possible to design of a 3D immersive platform to respond to the problem of designing, testing, and evaluating immersive eco-driving applications (involving the human factor in the loop). In order to achieve this HIL (Hardware In the Loop) platform, two technical tasks had to be addressed in Pro-SiVIC. The first was to adapt the consumption sensor that already existed. The second one required designing a new optical sensor for the real-time generation of a couple of images enabling the display of the 3D scene in the virtual reality helmet (first made with the OCULUS helmet, and then with the HTC Vive helmet) (Figure 4.42).

Figure 4.42. *Generation of a couple of HD images for 3D reproduction in the OCULUS and HTC Vive immersive VR helmets*

For this, a human stereoscopic vision (eye spacing 64 mm on average) with an adapted parallax was reproduced. In this 3D "*sensor*", the two images are generated in high resolution (1280x1024) and simultaneously sent as a single "*dual screen*" image to the helmet (Figure 4.43). This transfer is performed using the DDS communication bus. Using the IR sensors and the inertial unit present in the helmet, it is easy to determine the orientation of the driver's head. This information is then used in Pro-SiVIC so as to control the extrinsic configuration of the "*Dual Screen*" sensor. In this stereoscopic model, frequency and parallax can be adjusted during operation in order to optimize visual comfort for the real driver. Consumption and speed information in the form of gauges and meters embedded in the "*dual view*" plug-in are also available at an infinite distance (Figures 4.42 and 4.43). In order to increase the immersive realism of this platform, the driver has several experimental benches: a SECMA F16 vehicle, a Logitech cockpit, and a DEVELTER cockpit (Figure 4.45). In the SECMA vehicle, the speed and steering instructions are sent to Pro-SiVIC, via a CAN bus and RTMaps (Figure 4.44).

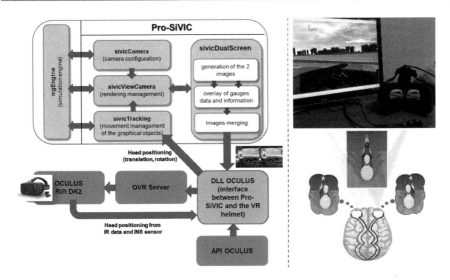

Figure 4.43. *Functional diagram: interfacing VR helmet with Pro-SiVIC*

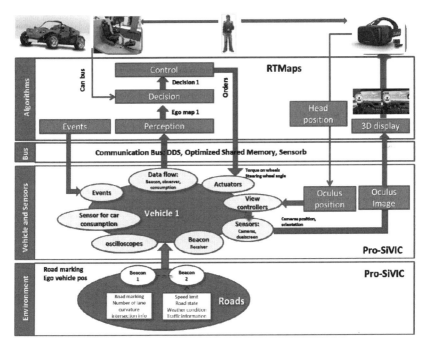

Figure 4.44. *Functional architecture of the "eco-SiVIC" application for prototyping, testing, and evaluating eco-mobility applications*

Figure 4.45. *Immersive eco-mobility experimental bench based on Pro-SiVIC (SECMA F16, Logitech driving seat, DEVELTER driving platform)*

The first research goals of this eco-mobility platform are to make drivers aware about eco-driving, and to be able to study their behavior in order to extract models, eco-responsible speed profiles (Orfila 2017a, 2019), and eco-driving advice (Geoffroy 2016). The scenario takes place on the virtual reproduction (digital twin) of the Satory test tracks.

In the experiments performed, the fuel tank was almost empty (0.15 liters of gasoline), and constraints were imposed on the drivers: the maximum speed was limited to 90 km/h, leaving the road was penalized (speed limit mechanism at 10 km/h). The experiments took place in three stages: first the driver had to drive normally, trying to cover the greatest distance. Then, he was given a piece of eco-driving advice (maintaining a constant speed between 40 and 50 km/h, limiting extreme accelerations/decelerations, avoiding steering wheel strong maneuvers), and the driver had to drive back, trying to cover the greatest distance in accordance with the eco-driving advice received. The ratio between the distance traveled and the optimum eco-driving distance provided a score.

4.7. Conclusion and perspectives

When we started working on the design of this simulation platform in 2002, the automotive industry was working on the preparation of new level 1 and 2 (SAE scale) driver assistance systems, leaving responsibility and driving decisions to the driver.

When in 2010, Google announced the design and deployment of vehicles equipped with automatic pilot systems assisted by a set of embedded sensors such as RADARs, video cameras and GPS circulating in California, the ambitions of automobile manufacturers and suppliers evolved significantly.

Since then, the industry has invested heavily in the development of systems that will eventually replace drivers (levels 3 to 5 automated driving systems). The underlying safety concerns therefore become essential and require implementing new methodologies in order to fully validate such new technologies (Gruyer 2017; Van Brummelen 2018).

Even if 93% of road accidents are due to errors involving the human factor, we should add that in 99.999% of critical driving situations, human beings manage to avoid the accident, by making a decision in line with the situation and environment.

The challenge is therefore launched: how can we certify the performance of new automated (levels 3 and 4) and autonomous (level 5) driving systems in order to move beyond the current performance of humans. A challenge of this certification is also to guarantee user safety for these new forms of mobility, under all circumstances.

We consider that simulation is becoming a candidate and an essential tool for certifying the performance of new perception systems (detection, monitoring, recognition), analysis and decision-making, and this, in all operating situations.

The work we have carried out on the simulation of sensors, vehicles and virtual environments for the prototyping, testing, evaluation and validation of ADAS – and more specifically, of embedded perception systems – is essential to enable us to industrialize a simulation solution, Pro-SiVIC, in line with the new needs of the automotive industry.

Indeed, since the launch of this research project in 2002, and more particularly since 2015 and the integration of simulation work within the French software publisher ESI, we have been able to work on improving realistic sensor models, and to compare them with real-life data from operating sensors. This comparison process was based on the use of our customers' use cases.

However, the relevance of technological choices had already been validated via a large number of collaborative projects (ARCOS, LOVe, ABV, HAVE-it, eFuture, CooPerCom, ISI-PADAS, Holides, etc.).

The Pro-SiVIC platform has also evolved acknowledging the industrialization of new standards, such as the FMI to enable the management of co-simulation, or even Open-Drive and Open-Scenario for creating 3D environment resources and road networks.

The challenges for the future will involve deciding how solutions like Pro-SiVIC should match verification, validation and certification processes/procedures for automated and connected vehicles. It is certainly regrettable to observe that so far, these procedures do not make it possible to profit effectively from the value offered by physico-realistic simulation.

Unfortunately, simulation is too often used as assistance during the advanced phases of system sizing or for verifying those systems, which functionally operate in defined and known fields. Indeed, the new decision systems based on a chain of perception such as camera, RADAR, LiDAR, etc. operate in open environments presenting infinite variability and complexity.

The future developments of simulation tools for the evaluation, validation and certification of ADAS for ACVs will involve extracting critical and rare cases – in which the system reaches the performance limit expected according to its design – and take into account the necessary certification-related decisions (before marketing), for convincing drivers to transfer the driving load to an automatic system.

It is certainly in this area that simulation will bring the most value to the automotive industry.

4.8. References

Bancroft, S. (1985). An algebraic solution of the GPS equations. *IEEE Transactions on Aerospace and Electronic Systems*, AES-21(1), 56–59.

Ben Jemaa, I., Gruyer, D., Glaser, S. (2016). Distributed simulation platform for cooperative ADAS testing and validation. *ITSC 2016*, Rio de Janeiro.

Blinn, J.F. (1976). Texture and reflection in computer generated images. *CACM*, 19(10), 542–547.

Blinn, J.F. (1977). Models of light reflection for computer synthesized pictures. *Proceedings of the 4th Annual Conference on Computer Graphics and Interactive Techniques (SIGGRAPH 77)*, 192–198.

Blinn, J.F. (1982). A generalization of algebraic surface drawing. *ACM Transactions on Graphics*, 1(3), 235–256.

Brown, D.C. (1966). Decentering distortion of lenses. *Photogrammetric Engineering*, 7, 444–462.

Choi, S. (2010). Estimation et contrôle pour le pilotage automatique de véhicule : stop&go et parking automatique. PhD Thesis, Mines ParisTech.

Choi, S., Boussard, C., d'Andréa-Novel, B. (2011). Easy path planning and robust control for automatic parallel parking. *IFAC 2011*, Milan.

Cohen M. and Wallace, J. (1993). *Radiosity and Realistic Image Synthesis*. Academic Press Professional, Cambridge.

Cook., L.R. and Torrance, K.E. (1982). A reflectance model for computer graphics. *ACM Transactions on Graphics (TOG)*, 1(1).

Demmel, S., Lambert, A., Gruyer, D., Larue, G., Rakotonirainy, A. (2014). IEEE 802.11p empirical performance model from evaluations on test track. *Journal of Networks*, 9(6).

eMotive. (2010). Projet FUI eMotive, livrable du lot 3. Simulateur de système de détection.

Geoffroy, D., Gruyer, D., Orfila, O., Glaser, S., Rakotonirainy, A., Vaezipour, A., Demmel, S. (2016). Immersive driving simulation architecture to support gamified eco-driving instructions. *23rd ITS World Congress*, Melbourne, 10–14.

Georges, G. (2016). Algorithmes de calcul de positions GNSS basés sur les méthodes des moindres carrés avancées. Thesis, University of Technology of Belfort-Montbeliard.

Grapinet, M., Desouza, P., Smal, J-C., Blosseville, J-M. (2012). Characterization and simulation of optical sensors. *Transport Research Arena – Europe 2012*, Athens.

Gruyer, D., Hiblot, N., Desouza, P., Sauer, H., Monnier, B. (2010). A new generic virtual platform for cameras modeling. *Proceeding of International Conference VISION 2010*, Montigny-le-Bretonneux.

Gruyer, D., Grapinet, M., Desouza, P. (2012). Modeling and validation of a new generic virtual optical sensor for ADAS prototyping. *IV 2012*, Alcalá de Henares.

Gruyer, D., Demmel, S., d'Andrea-Novel, B., Larue, G., Rakotonirainy, A. (2013). Simulating cooperative systems applications: A new complete architecture. *International Journal of Advanced Computer Science and Applications (IJACSA)*, 4(12), 171–180.

Gruyer, D., Choi, S., Boussard, C., d'Andrea Novel, B. (2014a). From virtual to reality, how to prototype, test and evaluate new ADAS: Application to automatic car parking. *IEEE IV201*, Dearborn, Michigan.

Gruyer, D., Orfila, O., Judalet, V., Glaser, S. (2014b). New 3D immersive platform dedicated to prototyping, test, evaluation and acceptability of eco-driving applications. *VISION 2014*, Versailles.

Gruyer, D., Magnier, V., Hamdi, K., Claussmann, L., Orfila, O., Rakotonirainy, A. (2017). Perception, information processing and modeling: Critical stages for autonomous driving applications. *Annual Reviews in Control*, 44, 323–341.

Gruyer, D., Glaser, S., Chapoul, J. (2018). Simulation platform for the prototyping, testing, and validation of cooperative intelligent transportation systems at component level. *2018 Australasian Road Safety Conference*, Sydney.

Hadj-Bachir, M. and de Souza, P. (2019a). LIDAR sensor simulation in adverse weather condition for driving assistance development [Online]. Available at: https://hal.archives-ouvertes.fr/hal-01998668/document.

Hadj-Bachir, M., Abenius, E., Kedzia, J-C., de Souza, P. (2019b). Full virtual ADAS testing. application to the typical emergency braking EuroNCAP scenario [Online]. Available at: https://hal.archives-ouvertes.fr/hal-02000567/document.

Kedzia, J-C., de Souza, P., Gruyer, D. (2016). Advanced RADAR sensors modeling for driving assistance systems testing. *EuCAP 2016*, Davos.

Lafortune, E. (1996). Mathematical models and Monte Carlo algorithms for physically based rendering. PhD Thesis, Katholieke Universiteit Leuven, Belgium.

Lafortune, E., Foo, S-C., Torrance, K.E, Greenberg, D. (1997). Non-linear approximation of reflectance functions. *Siggraph97*, Los Angeles, CA.

Menhour, L., d'Andréa-Novel, B., Fliess, M., Gruyer, D., Mounier, H. (2018). An efficient model-free setting for longitudinal and lateral vehicle control. Validation through the interconnected Pro-SiVIC/RTMaps prototyping platform. *IEEE Transactions on Intelligent Transportation Systems*, 19(2), 461–475.

Ngan, A., Durand, F., Matusik, W. (2005). Experimental analysis of BRDF models. *Eurographics Symposium on Rendering (EGSR2005)*, Konstanz.

Orfila, O., Gruyer, D., Judalet, V., Revilloud, M. (2015). Ecodriving performances of human drivers in a virtual and realistic world. *IEEE Intelligent Vehicles Symposium 2015 (IV 2015)*, Seoul.

Orfila, O., Glaser, S., Gruyer, D. (2017a). Safe and ecological speed profile planning algorithm for autonomous vehicles using a parametric multiobjective optimization procedure. *FAST-ZERO 2017, Nara Kasugano International Forum*, Nara, Japan, 18–22.

Orfila, O., Freitas Salgueiredo, C., Saint Pierre, G., Sun, H., Li, Y., Gruyer, D., Glaser, S. (2017b). Fast computing and approximate fuel consumption modeling for Internal Combustion Engine passenger cars. *Transportation Research Part D*, 50.

Orfila, O., Gruyer, D., Hamdi, K., Glaser, S. (2019). Safe and ecological speed profile planning algorithm for autonomous vehicles using a parametric multi-objective optimization. *International Journal of Automotive Engineering (IJAE)*, 10(1), 26–33.

Pechberti, S. and Gruyer, D. (2018). Method for simulating wave propagation; simulator, computer program and recording medium for implementing the method. Patent no: US 10,133,834 B2.

Pechberti, S., Gruyer, D., Vigneron, V. (2012). Radar simulation in SiVIC platform for transportation issues. Antenna and propagation channel modeling. *ITSC 2012*, Anchorage.

Pechberti, S., Gruyer, D., Vigneron, V. (2013). Optimized simulation architecture for multimodal radar modeling: Application to automotive driving assistance system. *IEEE ITSC 2013*, The Hague.

Rondinone, M., Maneros, J., Krajzewicz, D., Bauza, R., Cataldi, P., Hrizi, F., Gozalvez, J., Kumar, V., Röckl, M., Lin, L. *et al.* (2013). iTETRIS: A modular simulation platform for the large scale evaluation of cooperative ITS applications. *Simulation Modeling Practice and Theory*, 34(2013), 99–125.

SENSORIS (2015). Vehicle sensor data cloud ingestion interface specification (v2.0.2) [Online]. Available at: https://lts.cms.here.com/static-cloud-content/Company_Site/2015_06/Vehicle_Sensor_Data_Cloud_Ingestion_Interface_Specification.pdf.

Thandavarayan, G., Sepulcre, M., Gozalvez, J. (2020). Generation of cooperative perception messages for connected and automated vehicles. *IEEE Transactions on Vehicular Technology*, 69(12), 16336–16341.

Van Brummelen, J., O'Brien, M., Gruyer, D., Najjaran, H. (2018). Autonomous vehicle perception: The technology of today and tomorrow. *Transportation Research Part C: Emerging Technologies*, 89, 384–406 [Online]. Available at: https://doi.org/10.1016/j.trc.2018.02.012.

Vanholme, B. (2012). Highly automated driving on highways based on legal safety. PhD Thesis, Université d'Evry Val d'Essonne, Évry.

Vanholme, B., Gruyer, D., Lusetti, B., Glaser, S., Mammar, S. (2013). Highly automated driving on highways based on legal safety. *IEEE Transaction on Intelligent Transportation Systems*, 14(1), 333–347.

5

Standards for Cooperative Intelligent Transport Systems (C-ITS)

This chapter addresses the question of standardized technologies enabling the exchange of data between vehicles, other road users, road infrastructure, urban infrastructure, traffic control centers and services management platforms in the cloud. These technologies, standardized to meet European terms by ETSI, CEN and ISO, are known under the name "Cooperative ITS" (C-ITS) and bring together several technologies and functionalities for the transfer, organization, securing and processing of data. The best known are undoubtedly short-range localized communication technologies, based on a form of WiFi adapted to moving vehicles (ITS-G5, within the frequency band 5.9 GHz). These assist vehicles in communicating directly with one another and with the road infrastructure without the support of a telecommunications infrastructure (V2X). They are mainly used for applications related to road safety and ADAS will directly benefit from these. Cooperative ITS technologies also include long-range centralized communication technologies based on the cellular network (LTE, 5G), geo-localized data organization functionalities (Local Dynamic Map) and many others, which are being standardized. To ensure interoperability, these technologies are grouped together and integrated into a unified communication architecture (ITS station architecture) whose motivations, origins, use cases and vast sets of features will be further explained in this chapter.

Chapter written by Thierry ERNST.

5.1. Context and goals

5.1.1. *Intelligent transport systems (ITS)*

In the future, the mobility of goods and people will have to respond to major challenges, such as improving road safety (goal of the European Commission to drastically reduce the number of road accident victims), improving the efficiency of the road network in the face of the constant increase in road traffic, reducing the impact of mobility on the environment (pollution, space occupied by roads or parked vehicles) and many others.

To meet these challenges, it is essential to reliably and efficiently combine data drawn from very different environments, especially those produced by vehicle-embedded sensors or roadside infrastructure equipment distributed along paths.

Improvements are expected thanks to the use of Information and Communication Technologies (ICT). We speak of *Intelligent Transport Systems (ITS)* when these technologies are applied to transport.

Thanks to ICT and the services they enable, vehicles become part of a system in which different objects are connected and cooperate with one another. One can speak of intelligent mobility, because the information, the collection and exchange of data – as well as the optimized management of communication networks – make it possible to build use cases and to deploy new mobility services for goods (purchases and remote monitoring) and for people (multimodal transport, carpooling, car-sharing, self-service bicycles).

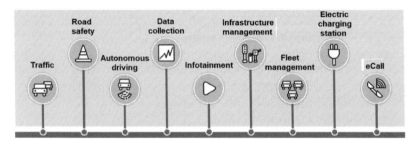

Figure 5.1. *Examples of ITS services. For a color version of this figure, see www.iste.co.uk/bensrhair/adas.zip*

Some of these services are shown in Figure 5.1. These new services can often be accessed via cloud platforms associated with geolocalization services (schedules, routes, finding a charging station or parking space, reservations, emergency calls, electronic tolls, etc.). Other services increasingly rely on vehicle sensors (cruise control, reversing radar, presence detection at blind spots, etc.) and on road side infrastructure (magnetic loop for detecting a vehicle at a traffic light, speed measurement camera, etc.).

These ever more numerous and connected services foreshadow the vehicle of the future, which will be more and more autonomous, but above all, connected and cooperative. Indeed, the latest advances in these technologies enable data exchange between vehicles, road and urban infrastructure, road users (drivers, passengers and vulnerable road users) and management and service platforms.

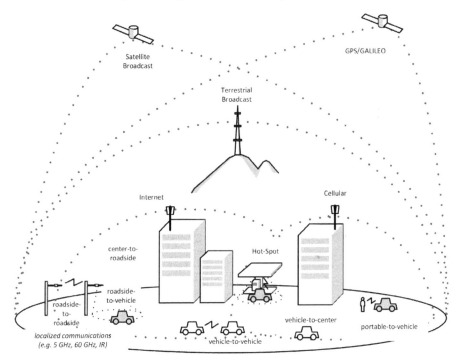

Figure 5.2. *Diversity of access technologies*

These data exchanges can take place via a *diversity of communication technologies* (Figure 5.2), in particular long-range communication technologies (cellular, satellite) which offer extended connectivity to the cloud, and short-range communication technologies (WiFi, infra-red), enabling vehicles to exchange data locally with their road environment (V2X).

5.1.2. *The connected and cooperative vehicle*

Latest generation vehicles are already connected objects. Connectivity makes it possible to provide a wide variety of telematic services (navigation, remote maintenance, emergency calls, fleet management and monitoring, etc.). Little by little, it enables the vehicle to become immersed in a road environment it can communicate with, thus improving the quality and reliability of the information available concerning vehicles, their location and the road environment. In the near future, vehicles will directly interact with one another and with road side and urban infrastructure equipment by means of localized communication technologies (V2X).

The new generations of vehicles are also equipped with driving assistance systems (ADAS) which assist the driver (in keeping distances, for changing lanes, etc.), thanks to the use of sensors. In the developed markets of Europe and North America, public authorities have made the use of driving assistance systems (Electronic Stability Control, ESC) mandatory for passenger and commercial vehicles since 2012. Also, the use of automated braking systems will become more and more widespread, something which will lead to a higher demand in vehicle and road side infrastructure technologies (radars, cameras and other sensors, communications).

These ADAS will continue to develop and will become cooperative, thanks to the integration of communications technologies. Sensors will be supplemented by precise localization and high definition mapping, or even coupled with sensor systems specially designed to match infrastructures. V2X communications will transmit information collected both from nearby vehicles and from the roadside infrastructure, and thus enrich the vehicle's perception functions at distances beyond the capacity of its own sensors (approximately 100 to 200 meters), in particular during unexpected traffic events (accidents, presence of pedestrians on the track, work areas, intervention of emergency vehicles, weather conditions).

Although this is still a distant reality, the combination of these technologies (sensors, ADAS, V2X communications, connectivity) will host the advent of the autonomous vehicle. For this, the contribution of communication technologies will be essential. Initially, they will make it possible to remotely take charge of the vehicle at low speed (remote control for parking in a car park, approaching an electric charging station, obstacle avoidance and safety along the roadside, etc.); this is already being done in the context of autonomous shuttle deployment. Then, in situations where driving delegation is authorized (motorway, automated parking, etc.), these will control whether traffic conditions are suitable for driving delegation functions, the presence of embedded high-definition maps and improve the vehicle's dynamic perception of the environment. Sooner or later, when the vehicle achieves full autonomy (level 5), these tools will enable convoy driving (synchronization of all the vehicles in the convoy to accelerate, slow down, allow for an insertion, etc.).

In any case, the fusion of data obtained from vehicle sensors, V2X communications and cellular connectivity will make the vehicle safer.

It is important to highlight that the factor helping to accelerate the deployment of these new technologies is the shift towards stricter regulations. In addition to being one of the main causes of road accidents, the lack of safety in road transport entails high costs and hinders overall economic development. Many initiatives are being undertaken to set up state-of-the-art traffic management systems and increase general awareness about the importance of road safety. To deal with the million annual deaths worldwide due to reckless driving accidents (McKinsey study 2018), governments are regulating or encouraging vehicle manufacturers to adopt certain automated safety features, such as the Autonomous Emergency Braking (AEB) system which will become mandatory by 2022 for all new cars in the United States. The serial adoption of this feature by a manufacturer awards them additional points on the Euro New Car Assessment (EuroNCAP) program in Europe. The integration of V2X communications systems as safety support is already underway in the 2024 protocol.

5.1.3. *Silos communication systems*

As we have seen in the previous sections, many ITS services are based on connectivity: electronic toll collection, navigation, fleet monitoring, traffic information, emergency calls (eCalls), road safety, etc.

Different communication solutions, often proprietary protocols, sometimes conforming to national or international standards, or based on *de facto* standards, are offered in accordance with the requirements of the service to be provided. A specific communication system is deployed for each type of service, or even by each stakeholder (each vehicle manufacturer, each fleet manager, each connected service provider, etc.).

This leads to the installation of *several vehicle communication systems*, all the more in specialized vehicles. Each system is based on tried and tested simple technologies, but which are very limited in their functionalities, their performance, and their safety levels. In addition, each of these communication systems is designed according to a selection of technologies which are specific for each system (radio technology, communication protocol, data format, server). This presents many drawbacks: lack of interoperability and complexity for exchanging data between systems and suppliers, no sharing of communication resources, non-optimized power consumption, ergonomic problems, high cost of communications (several SIM cards, several invoices), different and sometimes incompatible approaches (in particular concerning secure access to vehicle data and respect for personal data), etc.

It thus becomes obvious that, in the absence of a technological breakthrough enabling the convergence of the different communication systems, they tend to multiply in an anarchic manner.

5.1.4. Cooperative Intelligent Transport Systems (C-ITS)

New technologies called "Cooperative ITS" (C-ITS) have the particularity of enabling the exchange of data between road vehicles (private vehicles, utility vehicles, etc.), roadside and urban infrastructure (traffic lights, variable message signs, toll gates, etc.), nomad systems (smartphones, tablets, etc.) and cloud access servers (traffic control center, fleet management platform, database, etc.).

Not only do they improve the safety and efficiency of road traffic, but they can also provide new mobility services. Thus, they respond to the need for a *convergence of communication systems*, developed in different silos.

To guarantee the convergence and interoperability between stakeholders and solution providers, these new technologies are standardized and grouped together under a communication and data exchange architecture ("ITS station" architecture [ISO 21217], see International Organization for Standardization (2020), see International Organization for Standardization (2014)) which will be further developed in sections 5.2 and 5.3.

5.1.5. Diversity of Cooperative ITS services

Thanks to the unified "ITS station" communication architecture and the standardization of communication and data exchange functionalities, communication systems initially deployed for distinct use cases (for example, signaling and navigation) can exchange data and thus become "cooperative".

In this way, infrastructure operators will be able to collect a great deal of information emanating directly from equipped vehicles (type of vehicle, engine, axle weight, number of occupants, access authorizations, etc.). They will be able to offer new services to road users: information certified by road infrastructure operators (traffic conditions, roadwork in progress, danger zones, relief route, etc.), as well as value-added services (parking reservation, tourist information, traffic lights, reserved lane, etc.). Road infrastructure operators will be able to improve infrastructure maintenance (predictive maintenance and detection of road, bridge and tunnel defects) and make better use of their equipment (variable message signs, traffic lights, toll gates or parking facilities, etc.), and eventually enrich their use.

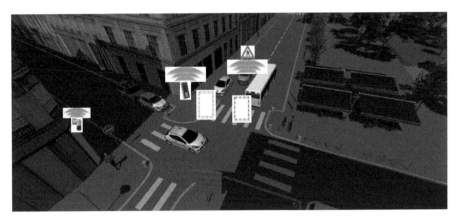

Figure 5.3. *Road safety improved thanks to communication between infrastructure and vehicles. For a color version of this figure, see www.iste.co.uk/bensrhair/adas.zip*

The technologies and standards of Cooperative ITS make it possible to collect the data available on the road environment (sensors of all types on the road side/urban infrastructure, or embedded into vehicles), to put them in order and transmit them to other stakeholders (other vehicles, other road users, road infrastructure, service providers, fleet managers, etc.).

Figure 5.4. *Time-critical data sharing between vehicles thanks to localized communications (1/2). For a color version of this figure, see www.iste.co.uk/bensrhair/adas.zip*

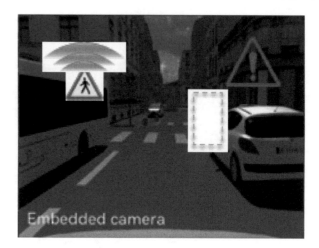

Figure 5.5. *Time-critical data sharing between vehicles thanks to localized communications (2/2). For a color version of this figure, see www.iste.co.uk/bensrhair/adas.zip*

Many applications are possible, in particular the exchange of data between vehicles and the road infrastructure so as to avoid accidents and optimize road traffic.

A first batch of C-ITS services (Day-1 services) is already the subject of pilot deployments throughout Europe (see section 5.5) (C-Roads 2019; InterCor 2021; Scoop 2019; Indid n/a). In addition to these, the innovative services below illustrate the range of use of the same technology as well as its features (genericity, openness, flexibility, interoperability, etc.):

– Services which improve road safety in urban areas (see Figures 5.3 to 5.5):

- detection by the bus or by the infrastructure of a pedestrian crossing the road in front of a bus, and dissemination of the information to vehicles approaching the bus for overtaking,

- detection by road infrastructure or by connected public lighting of a person crossing the road, light intensity variation and dissemination of the information to approaching vehicles,

- announcement of a slow-moving intervention-specialized vehicle (sweeper, collection of bulky items, etc.) enabling speed adaptation and selection of another route.

– Services enabling differentiated access to certain areas:

- detection by urban infrastructure equipment (retractable block, barrier) of vehicles and persons authorized to access a certain area depending on the time of day, the object (delivery), vehicle motorization or the number people transported,

- detection of a parking space and reservation, thanks to the use of sensors or cameras, and dissemination of information via standardized localized communications.

– Existing value-added services which can benefit from a shared communication platform offering several services:

- reservation and access to electric charging stations: localization, reservation, billing,

- reservation and access authentication for a car-sharing vehicle,

- automated coupling for a vehicle equipped with sensors enabling low-speed autonomous driving.

– New high value-added services not directly related to mobility, but requiring the inclusion of mobility data:

- air quality analysis,

- localized weather forecasts,

- predictive maintenance.

– Services enabling better information sharing between the different stakeholders involved with a construction site, a fire, an accident, an airport area, a warehouse, a mine, etc. in order to improve the safety of vulnerable people (pedestrians, cyclists, light vehicles), and to optimize travel expenses and operational costs.

5.1.6. *Standardization bodies*

The functionalities required for the deployment of "Cooperative ITS" services are specified by several standardization bodies, and more particularly in Europe (CEN, ETSI, ISO):

– The European Committee for Standardization (CEN, French: *Comité européen de normalisation*) and the International Organization for Standardization (ISO) are multi-disciplinary bodies whose members are the national standardization bodies (AFNOR, in the case of France). The work of CEN and ISO is monitored by a national standardization committee which represents the country at the international level. The technical committees developing standards in relation to Cooperative-ITS are ISO TC 204 ("ITS" technical committee), CEN TC 278 ("ITS" technical committee) and most recently CEN TC 226 ("road equipment technical committee"). These technical committees are structured into several working groups, each dealing with a specific aspect of ITS (emergency calls, public transport, electronic tolls, nomadic systems, communications, urban applications, etc.). The works of the working group dedicated to cooperative ITS (joint group ISO TC 204 WG16 and CEN TC 278 WG18) increasingly impact the work of other working groups. The standards produced by ISO and CEN are accessible for all, but they are not free.

– The European Telecommunications Standards Institute (ETSI) is a European organization specializing in telecommunications whose members are legal, industrial or academic persons. The "ITS" technical committee is dedicated to the development of V2X localized communications standards based on ITS-G5 radio technology in the 5.9 GHz band frequency. Unlike ISO and CEN, one has to pay to participate, depending on the size of the structure. However, published standards are accessible free of charge.

– The Institute for Electrical and Electronics Engineers (IEEE) and the Society of Automotive Engineers (SAE), for their part, are developing the V2X localized communications standards deployed in North America and a few other regions (IEEE P1609 WAVE/DSRC communication technology, equivalent to the European communication technology ITS-G5 from ETSI). The IEEE and SAE specifications do not employ the term "Cooperative ITS" and do not refer to the "ITS station" unified communication architecture described in the following sections. Some of these IEEE and SAE standards are also included in the European nomenclature. Participation in IEEE meetings is free of charge, and although accessible to all, standards must be paid for, as in the case of ISO.

5.1.7. *Genesis of the "Cooperative ITS" standards*

The need for communication architecture applicable to a variety of ITS use cases emerged as early as 2000, within ISO TC 204. Recognizing the difficulty of interoperability between systems each using a specific technology and communication protocols, in 2001 ISO TC 204 took the lead in creating a working group dedicated to communications (WG 16).

A communication architecture integrating several access technologies emerged from the very first meetings. These early architectural works are known under the name "CALM" (Communications Access for Land Mobiles), which was originally the name of the working group.

The first version of the architecture already integrates several access technologies for localized communications: WiFi variant dedicated to inter-vehicle communications in the 5.9 GHz frequency band (ISO 21215), millimeter waves in the 60 GHz frequency band (ISO 21216), and infra-red (ISO 21214). The same applies to centralized communications (satellite, cellular).

In 2006, several R&D projects claiming to comply with "Cooperative ITS" standards were launched within the framework of the 6th R&D program of the European Commission, in response to the call for projects devoted to "cooperative systems" (without specifically employing the term "ITS"). In particular, three flagship projects were launched: CVIS, SafeSpot, and Coopers. From 2008 onwards, smaller projects came to play with a focus on a specific technology (vehicular communications within GeoNet, secure communications within SeVeCom, road safety services within PreDrive C2X, etc.).

The CVIS (Cooperative Vehicle-Infrastructure System) project was the most resounding one. It made it possible to prove the concept of "CALM" communication standards for ISO TC 204, and to make them evolve thanks to the feedback and the participation of many new key stakeholders. The specificity of this project was to consider widely varied use cases ("road safety", "road traffic efficiency" and "value-added services"), requiring the use of a variety of access technologies.

At the same time, the SafeSpot project (mainly bringing together car manufacturers and equipment manufacturers) and the Coopers project (bringing together road infrastructure operators) focused on a set of use cases which is more specific to the primary concerns of each of these key stakeholder groups ("road safety" and "road traffic efficiency"). While the former considered localized communications, the latter tended to be interested in centralized communications.

In their first deliverables, these projects have proven the interest of developing a unified communication architecture, in order to ensure interoperability. COMeSafety, a new project (a "specific support action") aimed at harmonizing work and developing a summarizing document, brought together people working on these subjects within the framework of standardization, European projects, and industrial associations, such as the Car2Car Communication Consortium. The COMeSafety deliverable includes details about the services offered, as well as the communication technologies for meeting such needs. The fundamentals of "ITS station" communication architecture as we know it today are actually anchored in the diversity of technologies needed to meet all these needs.

The need to develop conformity test specifications, for which ETSI is competent, has led ISO TC 204 participants to propose the creation of an "ITS" technical committee within ETSI. Its creation was confirmed in 2008, and ETSI quickly specialized in the development of standards enabling the geo-localized data distribution between vehicles in the 5.9 GHz frequency band, reserved for ITS uses.

In 2009, the European Commission mandated ETSI and CEN to produce the standards needed for the rapid deployment of Cooperative ITS (Standardization Mandate M / 453, see European Commission (2019)). In 2010, this gave rise to the creation of the "Cooperative ITS" working group, joining efforts between ISO TC 204 (WG 18) and CEN TC 278 (WG 16) in order to develop the specifications of the necessary Cooperative ITS services, in addition to the works of ETSI TC ITS and ISO TC 204 WG16, specifically dealing with certain aspects of communications. In fact, the division of tasks between CEN and ETSI is not clear, because ISO TC 204 WG16 has been working since 2001 on the architecture and a variety of communication technologies, whereas since its inception ETSI CT ITS has been working only on Cooperative ITS services on the initiative of vehicles (CAM, DENM) using the ITS-G5 access technology. In 2012, the two groups from CEN and ETSI provided the European Commission with a report listing already developed standards, the ones currently under development and those to be developed [C-ITS Release 1] (see Intelligent Transport Systems (2013)). In particular, this report later allowed the identification of the priorities and missing standards, as well as the release of funding from the European Commission for the development of new standards by dedicated teams of experts: "Specialist Task Forces" (STF) at ETSI and "Project Teams" at CEN.

The CEN/ETSI report [C-ITS Release 1] (Intelligent Transport Systems 2013) dating from 2013 and, most of all, the explanatory brochure of the Cooperative ITS standards available on the CEN TC 278 website (C-ITS Brochure 2020), present a fairly exhaustive list of every C-ITS standard available at the time of publication.

5.2. "ITS station" architecture

5.2.1. *General description*

A standard presenting a "unified" communication architecture enabling data transmission between vehicles, road infrastructure and control centers, and potentially other entities, was effectively introduced in 2010 at ETSI (EN 302 665) and ISO [ISO 21217] (International Organization for Standardization 2020) based on the conclusions of the COMeSafety actions.

Originally identical, the ETSI version has remained fixed since its publication in 2010, whereas the ISO version has continuously evolved since its first edition, in order to clarify certain elements – mainly terminology-related – as well as to provide new functionalities – namely communications management as support for hybrid communications. The terminology is also available online [ITS-S Terminology] (see International Organization for Standardization 2014), but only in the 2014 version. It is therefore incomplete and we recommend obtaining the latest update of the 21217 standard (the 2020 version, or later).

The architecture, baptized as "*ITS station*" (ITS-S) is a "functional" type of architecture, and therefore deliberately abstract, in order to be applicable as widely as possible to all use cases, integration environments (vehicles, infrastructure, control centers, etc.), communication, security and data management protocols and technologies. Figure 5.6 shows the architecture in a commonly used simplified view. A somewhat more detailed, but still relatively abstract view is shown in Figure 5.7.

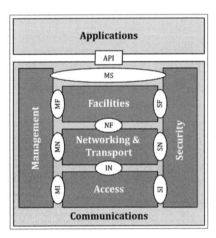

Figure 5.6. *Simplified diagram of the ITS station architecture (ISO 21217)*

The representation of functionalities follows the OSI model in seven superimposed layers commonly used for presenting the different functions which must be performed by a network-connected device, by isolating them from one another while interfacing the adjacent layers by means of Service Access Points (SAP).

In the "ITS station" model, the functionalities that carry out the functions of the OSI model are grouped into four horizontal layers, completed by two vertical entities which perform "inter-layer" functions:

– the "*access*" *layer* includes access technologies, for example, the protocols in charge of the radio aspects (ITS-G5, cellular, infra-red, etc.);

– the "*networking & transport*" *layer* includes network communication protocols (GeoNetworking, IPv6, 6LowPAN) as well as transport protocols (BTP, TCP/UDP, etc.);

– the *"facilities" layer* includes shared functions for data transmission, organization and fusion (V2X messaging, generic messaging, service publishing and discovery, LDM, PVT, etc.);

– the *"applications" layer* is devoted to the data processing functions specifically related to applications;

– the *"security" entity* reunites atomic security management functions which can be used by the four horizontal layers (management of secrets and certificates, authentication and encryption functions, etc.);

– the *"management" entity* provides management functions for the functionalities of each ITS station unit, for communications and for the lifecycle of the ITS station.

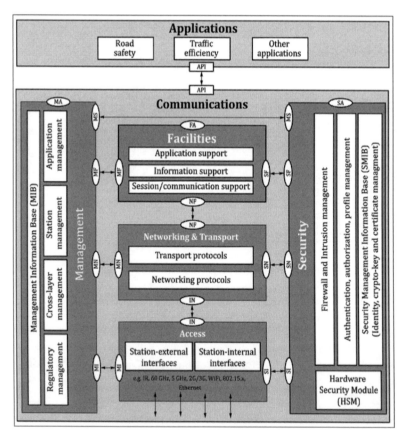

Figure 5.7. *Simplified diagram of the ITS station architecture (ISO 21217)*

As shown in Figures 5.6 and 5.7, the interfaces (SAP) between adjacent layers and entities are named after the initials of the pair of entities involved (e.g. "MN" for the SAP between the management entity and the networking and transport layer).

5.2.2. *ITS station communication units*

In no way do the standards impose a specific way of implementing the functions of the architecture. The various necessary functions can be implemented on a single device, or rather they can be dispersed into several communication units (*ITS station communication units* – ITS-SCU), and connected through an internal network (*ITS station internal network*). While these different communication units can work with a different subset of the architecture's functional blocks, they all include a "security" and a "management" entity, as illustrated in Figure 5.8:

– *ITS-S router*: this communication unit is dedicated to data transfer functions, and therefore involves radio signal routing and processing. This only calls forth the lower layers of the ITS station, as well as management and security functions.

– *ITS-S border router*: this communication unit is dedicated to the external routing function, that is to say, contacting the outside world with which the ITS station communicates. Compared to the basic ITS-S router, it has enhanced features so as to ensure security of communications and access to data (firewall, etc.). In a vehicle, localized communications using ITS-G5 or centralized communications using the cellular network are provided by the ITS-S border router.

– *ITS-S gateway*: this communication unit ensures the transfer of data between the ITS station and an attached sensor network (Sensor&Control Network – SCN), for example the CAN bus of a vehicle or a magnetic loop in the road infrastructure. These sensors – present on a potentially proprietary internal network – must not be directly accessible point-to-point from the outside. It is therefore necessary to go through a gateway at the application level, which ensures secured and differentiated access to the data available on the SCN.

– *ITS-S host*: this communication unit performs data processing functions, either shared (facilities layer), or specific to applications (applications layer).

5.2.3. *Types of ITS stations*

We traditionally speak of OBU (On-Board Unit), and more recently, of Vehicle *ITS station*, when referring to the equipment which integrate the functionalities enabling compliance with Cooperative ITS standards. As far as the roadside infrastructure is concerned, we traditionally speak of RSU (Road-Side Unit), and more recently, of *roadside ITS station*.

At present, Cooperative ITS standards specifically refer to four types of ITS stations (or "roles"), and to be strictly correct, one should say ITS Station Unit (ITS-SU).

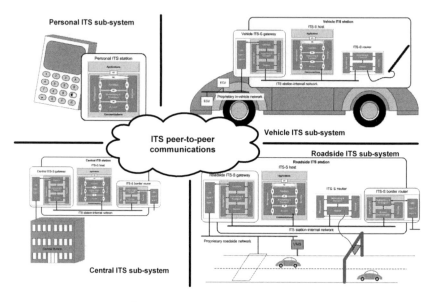

Figure 5.8. *Types of ITS stations*

Figure 5.8 illustrates these four types of ITS station units by showing an example on how to implement each of them:

– *Vehicle ITS station* (V-ITSS-S): when the equipment compliant with the ITS station architecture is deployed in vehicles, regardless of the type of vehicle and the mode of implementation (in one or more ITS-SCU communication units).

– *Roadside ITS station* (R-ITS-S): when the equipment compliant with the ITS station architecture is deployed in the roadside infrastructure (traffic lights, camera, variable message sign, etc.), without distinction of the mode of implementation (one could imagine that a 20 km stretch of road comprising several ITS-S routers installed on gantries, connected by optical fiber and controlled by an ITS-S host (a PC) only constitutes a single "ITS Station Unit").

– *Central ITS station* (C-ITS-S): when the services compliant with the ITS station architecture are deployed in the cloud (traffic control center, certification authority, vehicle fleet management platform, etc.).

– *Personal ITS station* (P-ITS-S): when the services compliant with the ITS station architecture are deployed on a mobile device, such as a smartphone.

5.3. Features of the ITS station architecture

Given the difficulties of finding documentation on ISO standards and the current narrow knowledge about them, in this chapter we have decided to give further details concerning the functionalities specified in the ISO standards, as well as key points about the functionalities specified by ETSI. For a more complete picture, the documents [C-ITS Release 1] (Intelligent Transport Systems 2013), [ITS-S Terminology] (International Organization for Standardization 2014) and (C-ITS Brochure 2020) prove to be a good reference source for the standards mentioned in this chapter.

5.3.1. *Combination of communication technologies*

"Cooperative ITS" standards make it possible to define services that do not depend on a predefined radio technology. A technology that is currently relevant can thus be replaced by a new, more efficient one without completely calling into question existing deployments and enabling a smooth transition from one technology to another, while ensuring interoperability. However, it leaves room for each stakeholder, country or region to make an appropriate choice depending on the density of the population, the areas to be served as a priority, costs, etc. It offers the possibility of using one or the other – or even combining them – to ensure a maximum penetration rate of these new services, knowing how to exploit one or the other technology.

The exchange between the different stakeholders can be done by using a variety of communication technologies and in accordance with two communication modes which complement one another, with or without going through the core of the telecommunications network:

– In the first case, data exchange takes place via a telecommunications network (cellular, satellite, etc.), and we therefore speak of *centralized communications* (or *networked communications*). These technologies are used to connect road infrastructure equipment to their control center, or to ensure connectivity between vehicles and connected service platforms (navigation, logistics, fleet monitoring, electronic payment and other telematic services, etc.). It is therefore necessary for the vehicle to be within the radio coverage of the telecommunications network. This is not always the case, because the deployment of the network in a territory takes several years and never covers it in its entirety (white zones); in addition, the network may malfunction, or lack performance (gray areas, urban canyon, network load, etc.).

– In the second case, the exchange takes place by *localized communications*, directly between vehicles and their environment (other vehicles, road and urban infrastructure, pedestrians), and therefore without a telecommunications network.

We then commonly speak of *V2X communications*. It is necessary for the vehicle to be close to another vehicle, or to infrastructure equipment paired with the same radio technology.

5.3.2. *Centralized communications*

Since its very beginning, the ITS station architecture has integrated several "centralized communications" access technologies, in particular satellite and cellular technologies, regardless of their forms (3G/4G/5G). Cellular technologies have their own architecture, covering all the layers of communications, and in particular "access" and "networking & transport" layers.

ISO standards only settle for defining the integration of these technologies, using them as "channels", that is as access technologies barely for the "access" layer. A more detailed integration, requiring interaction with the management entity, has not been documented.

The ISO 17515-2 standard specifies the integration of the D2D mode (Device 2 Device) in LTE (4G), recently standardized by 3GPP. This mode (PC5) makes it possible to avoid going through the heart of the telecommunications network when two vehicles are served by the same relay antenna. This mode still requires support from the telecommunications infrastructure, though one can shorten the communications path from vehicle to vehicle, and therefore reduce end-to-end delays. However, this functionality requires updating the cellular network in order to deliver the D2D functionality which has a cost – and of course one must be in an area covered by the cellular network.

5.3.3. *Localized communications (V2X)*

Unlike centralized communications commonly used for classic telematic services – which require the vehicle to be connected to the cloud and for which one can be satisfied with a communication solution specific for each service provider, or even owner – localized data exchanges between vehicles and with the road infrastructure require a solution that guarantees interoperability anywhere and with everything it matches. Therefore, localized data exchanges must follow certain standards.

With this in mind, a frequency band (5.9 GHz) has been reserved for ITS services in North America, Europe and other regions.

Standardization bodies (IEEE, ISO and ETSI) have developed a short-range communication solution (a few hundred meters), adapted to the 5.9 GHz frequency.

WiFi radio technology has been adapted to moving vehicles. This is a simplification of the general use of WiFi, which has removed the association functions which generate delays when establishing communications. For the neophyte, we will talk of vehicular WiFi; whereas for the more informed public, we will refer to it as the IEEE 802.11p standard.

ISO TC 204 has therefore developed the ISO 21215 (M5) standard which offers a layer of abstraction for integrating 802.11p access technology into the "ITS station" communication architecture, while harmonizing regional deviations standardized by IEEE (P1609) and ETSI (ITS-G5):

– In Europe, ITS-G5 is the radio technology currently being deployed for V2X communications. This operates in the 5.9 GHz frequency band dedicated to ITS services. The ETSI standard determines the use of this frequency band, among which one channel (CCH) is reserved for time-critical road safety services. Other ETSI standards specify the data exchange protocols on the CCH, which are GeoNetworking (communication in "broadcast" mode with multi-hop propagation until the message reaches a determined geographical area) and certain V2X messages associated with this mode of propagation (CAM, DENM, CPM, etc.). The use of other channels employing other transmission modes, in particular IPv6, is not fully specified.

– In North America, V2X communications are deployed in the same 5.9 GHz frequency band, but in accordance with the IEEE P1609 standards (WAVE/DSRC/WSMP architecture), which precede those by ETSI. This technology is commonly called DSRC, not to be confused with the European DSRC, which is a standard developed by CEN for electronic toll collection in the 5.8 GHz frequency band (RFID technology).

Localized communications can be performed thanks to other technologies, already integrated in the ITS station: millimeter waves in the 60 GHz frequency band (ISO 21216), infra-red (ISO 21214) and optical communications (ISO 22738). Other technologies are also under study, including:

– A new technology derived from 5G enabling direct vehicle-to-vehicle communications without going through the telecommunications infrastructure (Cellular V2X, developed by the 3GPP) is also arousing a lot of interest, but unlike ITS-G5/DSRC, it has not reached the level of maturity required for being broadly deployed. Indeed, several steps are still necessary before considering deployment: it is necessary to finalize the standards, to experiment with the technology on a large scale, to develop the conformity tests making it possible to validate the conformity of the solutions in relation to the standards, and above all, to have a dedicated frequency band.

– Finally, the P1609 group at IEEE started working on the evolution of vehicular WiFi to integrate the latest technological advances. Unlike Cellular V2X, this technology would be compatible with the ITS-G5 and P1609 technologies currently deployed in the 5.9 GHz frequency band in these regions.

5.3.4. *Hybrid communications*

The media talk very confusedly about the different forms of technology, without taking into account their distinct characteristics (use of a telecom infrastructure or not) or their level of maturity. In our opinion, technologies are complementary with one another and will be deployed at different rates from region to region, whereas other technologies – better suited for certain situations – will regularly appear, such as infrared or optical communications (LiFi/VLC). It is therefore necessary to combine these technologies intelligently in order to ensure interoperability, the transition from one technology to another and to provide more extensive connectivity.

It is of course possible to use different transmission modes and technologies simultaneously, in which case we speak of *hybrid communications*:

– for the same service in different situations depending on whether a vehicle is equipped or not or whether it is in a certain radio coverage: the same service can be given using localized or centralized communications – although in a degraded mode for certain cases, but it is preferable to not have any service at all;

– for distinct services or data flows, having different communication characteristics: for each type of service, one communication mode is more relevant than another; a vehicle equipped with both modes will be able to meet all the needs.

Thus, at a given moment, in addition to being connected to the cloud, each vehicle can also be equipped with a V2X communication system enabling the *localized exchange of data with other vehicles and with its environment* (road infrastructure, and eventually with other road users and urban infrastructure).

By way of example, an alert service to improve road safety could be carried out in an optimized manner using V2X localized communications (with ITS-G5) on major road routes, and centralized communications in rural areas (with the cellular network). Conversely, an area not served by the cellular network may give rise to the deployment of ITS-G5 equipment at the roadside, in order to ensure service coverage.

Figure 5.9 illustrates the profit of combining multiple communication technologies. Vehicle A (blue) detects a patch of ice. This information is of great use when related to other vehicles traveling on the same road, so that they slow down before they find themselves in the icy area. It represents time-critical information for vehicles following the blue vehicle; the information should reach them immediately, if possible by the use of localized communications (for example, a DENM message), if they are so equipped and within radio range. However, this information is still relevant for a certain time frame for all the other vehicles which will later travel on the same road section. It is essential for all the vehicles concerned to be informed at the right time, and this can only be done by using centralized communications, by informing a traffic control center which afterwards relays this information either via the cellular network (this requires a subscription for vehicles traveling in a specific area) or via roadside ITS stations (R-ITSS-S/RSU) located upstream of the area, which ensures transmission – provided that the vehicles are equipped. In a mountainous area, for example, it is quite possible to consider deploying RSUs at strategic crossroads.

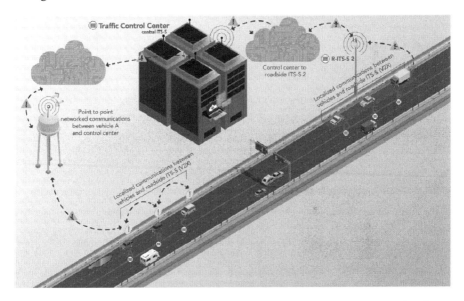

Figure 5.9. *Use of hybrid communications for the notification of hazardous situations. For a color version of this figure, see www.iste.co.uk/bensrhair/adas.zip*

From this example it is clear that it is difficult to ensure the transmission of the alert to all the vehicles by the use of a single type of technology. To achieve this, localized communications would have to be deployed on all vehicles and in all geographic areas, or the cellular network would have to be available everywhere, which will never be the case (whether in 3G, 4G or 5G). Combining technologies makes it possible to maximize the transmission of alerts while minimizing the cost of deploying a connected road infrastructure, or a cellular network in low-density or remote areas. It also makes it possible to organize a smooth transition from one technology to another (of different generations or different characteristics).

This technology-combining *hybrid approach* simplifies the development of standards and solutions and makes them sustainable, not only offering lower costs but, above all, enabling the development of innovative services based on data sharing.

5.3.5. *Extensive communications*

The use of several communication technologies simultaneously provides extended connectivity, dynamically exploiting all the access technologies available around the vehicle: connectivity is therefore more resilient compared to the use of a single access technology.

Apart from the fact that this maximizes the possibility of exchanging with any other entity (inside the vehicle with sensors, passengers and goods, other vehicles, other road users such as pedestrians or cyclists, sensor in the roadway, road or urban infrastructure equipment), above all, this lets the connectivity to the cloud be maintained as effectively as possible, anywhere and at any time, which is essential for a growing number of use cases.

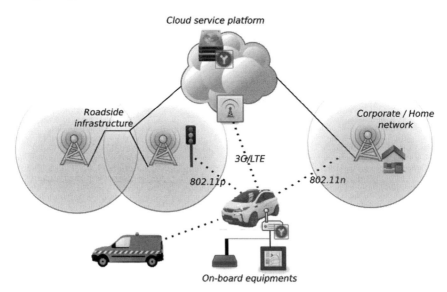

Figure 5.10. *Combination of multiple access technologies (localized and centralized communications) for providing extensive connectivity. For a color version of this figure, see www.iste.co.uk/bensrhair/adas.zip*

5.3.6. *Communications management*

Having several communication modes or technologies available requires the ability to choose the communication profile (list of protocols) best suited for each data flow.

To achieve this, cutting-edge functionalities have been integrated into the ITS station architecture in order to be able to dynamically select the protocol stack best suited for each data flow according to the access technologies actually available and their characteristics, on the basis of the communication constraints expressed by applications, which simultaneously contribute to resource consumption (CPU, bandwidth, etc.). A technical report (ISO 21186-2) expresses the main principles of the functionalities which are currently divided between several standards:

– The ISO 17423 standard lets application processes (ITS-S Application Process- ITS-S AP) define their constraints and needs in terms of communications for each data flow they will have to send: priority, transmission mode (broadcast or point-to-point), permanent connectivity, latency, data amount, cost, security level and services to be applied (confidentiality, time stamping, etc.). This can be done through communication parameters transmitted by the application to the "management" entity. The application can also impose the use of a set of predetermined protocols (communication profile). This may be necessary for very specific use cases, or to satisfy regulations, but is less flexible for cases when the profile is not applicable (e.g. lack of radio coverage of the technology requested).

– The ISO 24102-6 standard defines the generic functions required for meeting the constraints expressed by the applications. This is done by supervising the functionalities present in each layer of the ITS station (capabilities) and the environment of the ITS station, in order to determine which communication technologies are actually available and with what performance level (radio coverage, network load, etc.). A communication profile is chosen and instructions transmitted to each layer for ensuring coordination.

– The CEN 17496 standard specifies several communication profiles based on the concept of a unique identifier for finding the characteristics of each protocol in a register. This results in a common repository, facilitating interoperability. It also simplifies the drafting of technical specifications for C-ITS services. An earlier ISO version also exists (ISO 21185), the two of which will be combined in a future revision.

– The ISO 21210 standard defines the use of IPv6 in the context of communications between ITS stations. This protocol is essential for point-to-point or connected mode communications. IPv6 guarantees the sustainability of the telecommunications infrastructure and services in the face of the necessary transition of the Internet from IPv4 to IPv6 to overcome the lack of IP addresses. Knowing that IPv6 is not a protocol but a family of protocols, the ISO 21210 standard makes it possible to indicate the required IPv6 functionalities according to the type of ITS station, and also ensure interoperability between ITS stations; otherwise, different and incompatible choices could be made by different actors. For this purpose, this standard recommends a mechanism for transferring communications from one access technology to another, without interrupting current sessions (Mobile IPv6/NEMO).

5.3.7. *Messaging*

The available literature focuses almost exclusively on time-critical C-ITS services related to road safety. They are the most popular ones because they were involved in pilot deployments before other services (see section 5.5).

These time-critical services are based on localized communications (V2X) in broadcast mode and on the data transmitted by the following messages:

– *Cooperative Awareness Message (CAM)* (ETSI EN 302 637-2): this message allows a vehicle to make itself known to nearby ITS stations, in order to avoid collisions. For this, the message is broadcast by the vehicle at high frequency (up to ten times per second) and contains the vehicle's position, speed, acceleration, steering angle and other types of information.

– *Decentralized Environmental Notification Basic Service (DENM)* (EN 302 637-3): this message sent by a vehicle warns ITS stations against hazards on the road network, such as a patch of ice, an obstacle on the road, a stationary vehicle, etc.

– *Signal Phase and Time (SPAT) and Map Data (MAP)* (ISO 19091): this message emitted by the road infrastructure makes it possible to let other vehicles know the phase and duration of a stoplight, as well as the mapping of the intersection, in order to determine which lane each phase is applied to.

– *Cooperative Cooperation Message (CPM)* (work in progress): this message enables autonomous vehicles to exchange sensor information, in order to improve their perception at intersections.

These messages all contain a certificate which allows their sender to be authenticated while guaranteeing anonymity, through the use of pseudonyms (temporary identifiers avoid being able to associate the message with a specific vehicle or person).

However, it is important to distinguish other types of messages – and to define the following four messaging categories – in view of what has been discussed in the previous sections regarding localized, centralized and hybrid communications:

– C-ITS messaging based on localized communications (between vehicles and with the road infrastructure), according to a broadcast transmission mode (CAM, DEMN, CPM, SPAT/MAP): the standards in which these messages are currently specified have the disadvantage of being associated with a predetermined communication profile, in this case ITS-G5, GeoNetworking and BTP. The drawback of the current specifications is the inability to ensure the transmission of these messages using a different communication profile, without reviewing the standards.

– C-ITS messaging based on centralized communications (between vehicles and a control center), using a point-to-point transmission mode: as indicated in the previous point, the CAM, DEMN, CPM, SPAT/MAP messages are not suitable for this transmission mode. However, some deployments of C-ITS services attempt to transmit these same messages via the cellular network at the cost of interpretation of standards, or contortions which can undermine performance. The biggest drawback comes from the fact that the signing of messages is currently done directly at the GeoNetworking level, and that the latter needs to be encapsulated in an IP message in order to be transmitted across a cellular network.

– "History-based" messaging (not C-ITS) relies on a centralized transmission mode and specific communication protocols for each service. These messages were not designed for Cooperative ITS and do not comply with the ITS station architecture and its functionalities. However, the ITS station architecture ensures the continuity of these services by enabling the integration of "legacy applications" which settle for transmitting the data of these services without using the advanced functionalities of the ITS station, and therefore without benefiting from them.

– Generic C-ITS messaging which draws advantages from the advanced functions of the ITS station, in particular the management of hybrid communications, security, data processing (LDM, publish-subscribe, etc.) and other shared functionalities (facilities). This messaging service is defined in the ISO 17429 standard. This standard specifies a generic message header (Facilities Service Header – FSH) which precedes the data that can be added in cascade, and are organized according to the data dictionary in which they are defined (using the dictionary identifiers and elementary data). This messaging works in all communication modes (broadcast or point-to-point) and applies to all communication profiles thanks to the knowledge of the transmission constraints indicated by the application processes (ITS-S AP) using the ISO 17423 and ISO 24102-6 standards (see previous section).

These messaging services are made available for application services which can use them just the way they are, or supplement them, without having to develop an *ad hoc* standard. For scaling considerations and a sustainable development of services, a good practice is to use generic messaging and to publish the data formats used by each service, something which facilitates the reuse of identical formats for different services, in addition to ensuring interoperability.

5.3.8. *Data organization and identification*

Many services or use cases use similar data. So far, the tendency has been to define its own format for each service – including for the position and date/time stamp data commonly used in most services, which leads to incompatibilities, complexity and unnecessary effort. It is therefore necessary to define a data dictionary which contains the definitions of elementary data (in ASN.1), and to assign a globally unique identifier in the data dictionary. The standardization work contemplates the specification of a data management model enabling the creation of data dictionaries specific for each application field, or for each standardization body.

Data dictionaries can contain data formats whose description is published, or proprietary formats (format unknown to the public, or the community outside the one that defined it). However, for each data format, it is necessary to make a registration request so as to obtain a globally unique identifier. The process is currently under discussion at ISO and IEEE, and covers the identifiers of applications, messages, data and, to a lesser extent, those of protocols and facilities (their identification is necessary to define the communication profiles and the ability of the ITS to satisfy them).

Thanks to the concept of dictionaries, it becomes easy for each new service to retrieve data formats published in the dictionary, and to refer to them, rather than redefining a format again. Naturally, there are situations in which the published data formats do not meet a particular requirement, in which case it suffices to define a new format, and to proceed to its registration, for the benefit of the whole community.

This approach, still under construction in terms of standards, is related to two essential above-mentioned functionalities:

– The *geolocalized database* (Local Dynamic Map – *LDM*) (ISO 18750/ETSI EN 302 895): the data collected by an ITS station and stored in the facilities layer with attributes for identifying the date and place of relevance of these data. The recorded data are more or less dynamic (nearby vehicle positions, obstacles on the roads, roadwork areas, remarkable points). The LDM is used by every C-ITS service described in the previous section.

– A data *publication-subscription* mechanism: lets an application process (ITS-S AP) publish the data at its disposal (something which could be done by a sensor, or by a fusion algorithm creating new merged data from those obtained as input, for example) in order to enrich the knowledge base and make other application processes benefit from the same ITS station, or even share them with neighboring or remote ITS stations, by means of a messaging service. Such a mechanism is necessary to access data stored in the LDM (ISO 21184) or processed by messaging (ISO 17429).

5.3.9. *Secure communications and access to data*

Communication and data security is a central element in ITS station architecture. The basic principle is that an ITS station is a trust domain (BSMD: Bounded – Secured – Managed – Domain). The standards do not define the modalities making it possible to ensure that an ITS station implementation respects this rule, since this is specific for each integration environment, for each operating system and for the organization between different communication units (ITS-SCU).

However, the functions needed for securing communications and data are grouped under the "security" entity, in order to be accessible to all the functions transmitting or receiving data (in ISO jargon, we speak of application processes – ITS-S AP), and these can be found in all the layers or entities of the ITS station since they may well refer to application services, to communication services management or to the lifecycle of the ITS station.

The "security" entity includes functions for gaining access to certificates obtained from a certification authority, and based on a public-private key mechanism. The certificates are used in the exchanges between ITS stations in order to ensure the authentication of the transmitter with the receiver, whether in "broadcast" or in "point-to-point session" mode:

– in the first case (C-ITS messaging based on V2X localized communications), the content of the message is visible to everyone. The IEEE P1609.2 certificate guarantees that the content has reached the receiver without modification, and that the transmitter is legitimate in the role claimed. The messages are signed at the level of the "network" layer (GeoNetworking);

– in the second case, secure sessions are organized between ITS stations (ISO 21177), using IEEE P1609.2 certificates and the TLS 1.3 protocol (RFC 8446 with extension). Among other things, the certificate makes it possible to identify the transmission application process (ITS-S AP).

Another principle is that application processes can only access data matching their role and the permissions granted to them. This verification can be carried out by presenting the certificates.

5.3.10. *Evolution of standards*

New uses are continually emerging. They may have specific needs, and lead to the development of new services, to the definition of data formats and new messages meeting the specific needs of each use case. New access technologies are also constantly appearing, and existing technologies popular in other fields could usefully be integrated into ITS station architecture.

Fortunately, the ITS station architecture enables the integration of new functionalities.

The work required in the short and medium term, some of which is already underway, will focus on:

– the integration of new access technologies (5G, Cellular V2X, LiFi, LoRa, Bluetooth, etc.) and sensor networks;

– data organization functionalities: formal definition of data dictionaries, data formats and an identifier registry (ISO 5345 and ISO 5146 standards are under development);

– the definition of a new generation of services taking advantage of data organization into dictionaries and the management of hybrid communications;

– periodic standard review, taking into account feedback from pilot deployments throughout Europe.

5.4. Features of the ITS station architecture

The ITS station architecture optimally combines different technological bricks within a unified architecture for all connected & cooperative mobility uses. The principle of a modular architecture with different layers independent from one another, as well as supplemented by transversal management and security functions, has many advantages:

– consistently organizing every form of communication concerning the vehicle, its equipment, its users, or its peers (other vehicles, road infrastructure). These communications can either be time-critical messages related to road safety or exchanges requiring a high (cyber-)security level, but without constraints on latency. This harmonious organization reduces the costs related to network communications or to the multiplicity of equipment, facilitates the development and deployment of new applications within the vehicle, and allows for use cases involving vehicle equipment without compromising the vehicle's safety;

– performing both end-to-end communications between remote machines using the telecommunications infrastructure (centralized communications) and direct exchanges, without the support of a fixed telecommunications infrastructure (localized communications);

– managing communications flow by flow: each data flow can be prioritized, processed according to its characteristics and oriented towards the most appropriate access technology;

– combining ITS-specific technologies (including V2X localized communications, LDM) with traditional Internet technologies (IPv6, security, service discovery, etc.);

– having shared services (facilities) which simplify the development of application services as well as data sharing and fusion by using a data dictionary in which the elements are identifiable, together with publication, recording and subscription functionalities.

5.5. Deployment of Cooperative ITS services

Services based on *localized communications (V2X) require a coordinated deployment of interoperable solutions*, for the road infrastructure and for vehicles, from different manufacturers. Indeed, during the first years of deployment, it is unlikely that two equipped vehicles will meet in a situation where a useful event is generated. To speed up deployment, it is necessary to equip vehicles from captive fleets to circulate at a specific area and to simultaneously deploy equipment in the road infrastructure (RSU/roadside ITS station) in notable places where the first equipped vehicle is able to benefit from the technology. This is what led the French authorities to deploy C-ITS technologies on major highways, giving priority to the road operators' fleet of vehicles (several hundred vehicles, in particular patrol vehicles).

The first experiments, carried out across Europe, were concluded in 2013 and 2014. Their initial goal was to validate the concept of V2X communications for road-safety related uses. This was prominently the case of the SCOREF project in France; at the European level with DriveC2X and FOTSIS projects and especially in the United States, in Ann Arbor (Michigan), where more than 2000 vehicles took part in such experiments.

Following these experiments, large-scale pilot deployments (several thousand vehicles and a few hundred communicating road infrastructure equipment for each of them), were launched in several European countries (including SCOOP in France, ITS Corridor in Germany, eCoAT in Austria, NordicWay in the Scandinavian countries) in order to validate the interest of the technology with professional users in real conditions of use, and above all, in order to implement the entire ecosystem of stakeholders prior to a massive deployment.

These pilot deployments are all based on the same C-ITS standards set, although there are discrepancies on the versions used (not all deployments were started at the same time) and local preferences on priority services and technologies (some countries preferred to deploy services based on V2X localized communications with ITS-G5, others on centralized communications with LTE, or a hybridization of the two).

These deployments are harmonized; priority services have been defined since 2017 within the framework of the "C-ITS platform" of the European Commission [C-ITS Platform] (see European Commission (2017)), and implemented in the European pilots. C-Roads (2019) is more oriented towards road safety services whereas InterCor (2021) is focused on logistics-oriented services.

In France, Mr. Cuvillier, the Minister of Transport who was in office in 2014, announced the launch of the French pilot deployment (Scoop 2019) on the occasion of Mobility 2.0 day in February 2014, and the inauguration of the VEDECOM Institute. The original plan was to deploy 200 road infrastructure equipment and 3,000 vehicles (1,000 vehicles from public and private road infrastructure operators, in particular DIR Ile de France, DIR Ouest, the DIR Atlantique, SANEF and Vinci; a series of 1,000 vehicles sold by Renault and just as many by PSA). The first calls for tenders to equip the road infrastructure and the vehicles of the road infrastructure operators were published in mid-2016 and continued until 2019. Even though the target of 3,000 vehicles was not met, the number of units effectively deployed in 2020 as part of the SCOOP pilot makes France the deployment leader in Europe.

InDiD (*Infrastructure digitale de demain*) (Indid n/a), a new deployment program following SCOOP, was launched at the end of 2019 with a budget of € 10 million. Spanning five years and coordinated by the Ministry of Ecological and Solidarity Transition, it brings together 25 partners including communities (the Metropolis of Grenoble, the City of Paris, the Metropolis of Aix-Marseille and the districts of Isère and Bouches-du-Rhône), road infrastructure operators (DIRs, APRR, Vinci, Atlandes, SANEF and ASFA) and user communities (Valeo, Renault, PSA, TomTom). New public contracts were launched in 2019 (DIR Est, DIR Nord) and others are in preparation.

Large-scale deployment should take place on a massive scale from 2023 and may be imposed by the European Commission. Indeed, the imminent vote of a Delegated Act setting the conditions of the deployment was announced on March 13, 2019. In addition to the 2010 ITS directive, the text highlights the ITS-G5 technology and 3G/4G cellular networks (qualified as "mature" technologies), paving the way for future developments (LTE-V2X and 5G). The text was to be voted on by parliamentarians in the summer of 2019 [C-ITS Directive] (see European Commission (2019b)) but the vote was postponed *sine die*, without calling into question the deployments in progress, in particular those carried out on the road infrastructure as part of the C-Roads deployment program, which is being carried out smoothly. Although not essential for massive deployment, the Directive's project should return to the foreground in 2021. Above all, its ratification will make it possible to establish a framework ensuring interoperability and encourage those who are still hesitant to embark on deployment.

It should be noted that Australia and Israel have also decided to deploy C-ITS services in accordance with European standards. European C-ITS standards are viewed as more flexible than their North American equivalent, since deployment on the basis of European standards can be done alternately by using localized V2X (ITS-G5) or centralized (cellular) communication technologies. For Australia, the choice fell on the flexibility of the European approach given the extent of the territory which pushes to deploy C-ITS services by using a hybridization of technologies (ITS-G5 on certain axes, cellular-based services elsewhere).

"Hybrid" solutions are effectively implemented as part of C-ITS services pilot deployments in Europe: in Norway (NordicWay), C-ITS services are primarily delivered by the cellular network, whereas in France (SCOOP), they are primarily delivered with ITS-G5. In addition to facilitating and accelerating the deployment, this hybrid approach could satisfy all the stakeholders, on the condition that interoperability is ensured.

As for manufacturers, from just a few thousand units at present, the number of vehicles equipped with V2X technologies is estimated at several million within a few years. Initiatives issuing from car manufacturers have been announced as regards the deployment of V2X services:

– Volkswagen has been marketing a new series of its "Golf" model since the end of 2019. This integrates V2X services based on the ITS-G5 technology for improving road safety, which is justified by the EuroNCAP program. Volkswagen includes this technology, at no cost to the purchaser of the vehicle, as a standard function in order to accelerate the penetration of V2X technology in Europe.

– Since 2017, in the United States, Cadillac has deployed V2X communication technology based on the IEEE P1609 (DSRC) standards on its Sedan CTS model.

The following year, the car manufacturer announced its plan to expand the technology within its portfolio of new vehicles by 2023. To date, more than 100,000 Toyota and Lexus vehicles equipped with V2X have been sold. The Hyundai manufacturer has partnered with Autotalks in order to deploy V2X in its Genesys G90 model by 2021.

However, all manufacturers do not necessarily rely on the deployment of localized communication technologies (ITS-G5/DSRC). Some wish to carry out vehicle-to-vehicle data exchanges using the cellular network (5G), and therefore resort to the telecom infrastructure. This presents limitations for road safety applications, mainly due to data transmission latency and the uneven availability of the cellular network.

Finally, numerous experiments involving autonomous shuttles (Navya, EasyMile, Milla) use V2X communication technologies (SPaT/MAP and CPM messages), in order to secure the traffic light crossroads present on the experimental sites at the open road.

In conclusion, many use cases of connected vehicles which cooperate with their environment are based on the ability to exchange data between various entities, in various situations, with specific requirements for each type of communication flow (priority, transmission delay and latency, safety level, amount of data, continuity of service, etc.). These requirements cannot be met by just one type of communication technology; they must be intelligently combined within a standardized architecture to ensure interoperability. "Cooperative ITS" services – which are characterized by data exchanges between vehicles and other road users, roadside and urban infrastructure and cloud service centers – are based on the communication architecture of ITS stations and their corresponding data, security and management functionalities. These functionalities are standardized by ISO TC 204, CEN TC 278 and ETSI TC ITS. Thanks to these standards, the connected and cooperative mobility market is about to accelerate sharply. This acceleration follows the pilot deployments initiated throughout Europe in order to confirm the maturity of the underlying communication standards and the interest of Cooperative ITS services. In the long term, this deployment will affect the entire vehicle fleet and the roadside and urban infrastructure (traffic lights, variable message signs, car parks, electric charging stations, street furniture, etc.). These standards will prompt the deployment *in fine* of the connected and cooperative autonomous vehicle, and in particular, localized communications (V2X), which will gradually become integrated into driving assistance systems (ADAS).

5.6. References

C-ITS Brochure (2020). Cooperative Intelligent Transport Systems (C-ITS) Guidelines on the usage of standards, June [Online]. Available at: https://www.itsstandards.eu/app/uploads/sites/14/2020/10/C-ITS-Brochure-2020-FINAL.pdf.

C-Roads (2019). C-Roads – The platform of harmonized C-ITS deployment in Europe [Online]. Available at: http://www.c-roads.eu.

European Commission (2009). Standardisation mandate addressed to CEN, CENELEC and ETSI in the field of information and communication technologies to support the interoperability of co-operative systems for intelligent transport in the European community [Online]. Available at: https://ec.europa.eu/growth/tools-databases/mandates/index.cfm?fuseaction=search.detail&id=434.

European Commission (2017). C-ITS platform phase II: Cooperative Intelligent Transport Systems towards Cooperative, Connected and Automated Mobility. Final report [Online]. Available at: https://ec.europa.eu/transport/sites/transport/files/2017-09-c-its-platform-final-report.pdf or https://ec.europa.eu/transport/themes/its/c-its_en.

European Commission (2019a). EU Road Safety Policy Framework 2021–2030 – Next steps towards "Vision Zero". Staff working paper, SWD(2019), 283 final [Online]. Available at: https://ec.europa.eu/transport/sites/transport/files/legislation/swd20190283-roadsafety-vision-zero.pdf.

European Commission (2019b). Specification for the provision of cooperative intelligent transport systems (C-ITS). Draft delegated regulation [Online]. Available at: https://ec.europa.eu/info/law/better-regulation/initiatives/ares-2017-2592333_en.

European Commission (2020). Progress and findings in the harmonisation of EU-US security and communications standards in the field of cooperative systems: EU-US task force – Reports from HTG1 and HTG3 [Online]. Available at: https://ec.europa.eu/digital-single-market/en/news/progress-and-findings-harmonisation-eu-us-security-and-communications-standards-field [Accessed December 2020].

Indid (n/a). Infrastructure Digitale de Demain. Deployment plan of Cooperative ITS services in France, 2018. "Connecting Europe Facility" (CEF) Project, financed under European Grant number INEA/CEF/TRAN/M2018/1788494. Available at: https://ec.europa.eu/inea/en/connecting-europe-facility/cef-transport/2018-fr-tm-0097-s.

Intelligent Transport Systems (2013). Final joint CEN/ETSI-Progress Report to the European Commission on Mandate M/453. CEN TC ITS & ETSI TC ITS [Online]. Available at: https://www.etsi.org/images/files/technologies/Final_Joint_Mandate_M453_Report_2013-07-15.pdf [Accessed December 2020].

Intelligent Transport Systems (2020a). C-ITS Secure Communications. CEN TC278 Project Team PT1605 official web page [Online]. Available at: https://www. itsstandards.eu/highlighted-projects/c-its-secure-communications [Accessed December 2020].

Intelligent Transport Systems (2020b). C-ITS Secure Communications. CEN TC278 Project Team PT1605 annex web page [Online]. Available at: http://its-standards.eu/PTs/PT1605 [Accessed December 2020].

InterCor (2021). Interoperable Corridors deploying cooperative intelligent transport systems [Online]. Available at: https://intercor-project.eu [Accessed December 2020].

International Organization for Standardization (2014). Terminology extracted from standard ISO 21217, 3rd edition [Online] Available at: https://www.iso.org/obp/ui/#iso:std:iso:21217:ed-3:v1:en [Accessed December 2020].

International Organization for Standardization (2020). ITS station communication architecture for Intelligent Transport Systems. ISO TC 204. International Standard ISO number 21217, last edition 2020 [Online]. Available at: https://www.iso.org/standard/80257.html.

Scoop (2019). French pilot deployment of Cooperarive ITS Services 2015–2019 [Online]. Available at: http://www.scoop.developpement-durable.gouv.fr.

6

The Integration of Pedestrian Orientation for the Benefit of ADAS: A Moroccan Case Study

6.1. Introduction

Metaphorically speaking about road accidents as a "road war" is neither exaggerated nor arbitrary, since such accidents are among the leading causes of death for thousands of people every year.

It turns out that the term "driving" does not only closely refer to driving a vehicle, but also to social driving. It is through the latter that we will present the current state of road safety in Morocco. The situation is certainly chaotic, given the number of accidents which occur in the country. This is shown, for example, by the results of a survey of road traffic accidents and victims in Morocco in 2011: there was a 2.48% increase in accidents, which rose from 65,461 to 67,082 accidents, with 11.75% fatalities (rising from 3,778 to 4,222), 9.36% in serious injuries (which rose from 11,414 to 12,482) and 2.84% in minor injuries (which rose from 87,058 to 89,529). This alarming record [MOU 12] indicates that traffic accidents occur everywhere in Morocco, not only "in the cities, but also on sidewalks, in the countryside, on highways, on national, secondary, rural roads, etc."

Recognizing that behavior is a determining factor in traffic accidents raises technical and human scientists to the rank of experts; these fields have often been ignored in favor of technicist paradigms, leaving no room for questioning via

Chapter wirtten by Aouatif AMINE, Abdelaziz BENSRHAIR, Safaa DAFRALLAH and Stéphane MOUSSET.

For a color version of all the figures in this chapter, see www.iste.co.uk/bensrhair/adas.zip.

these same sciences, which are able to inform us with regard to behavior. Without underestimating the significance of factors related to roadside infrastructure (the condition of roads and vehicles, for example) in provoking accidents, reflections must now focus seriously on the human factor (human behavior: both that of drivers and pedestrians, as well as other vulnerable users). A high proportion of drivers break the Highway Code, drive in a drowsy state, have delusional beliefs in their power to control every situation – even while intoxicated – thus underestimating risks and acting unconsciously.

In this chapter, we will present a study aimed at helping the driver while they are driving and for improving pedestrian safety. This study was carried out as part of a road safety project in Morocco and was financed by the Ministry of Equipment, Transport, Logistics and Water (METLE, *Ministère d'Équipement, Transport, Logistique et de l'Eau*), in collaboration with the National Center for Scientific and Technical Research (CNRST, *Centre National pour la Recherche Scientifique et Technique*). In order to reduce fatal accidents involving pedestrians, much research is being carried out regarding ADAS (Advanced Driver Assistance System), specifically for the mitigation of pedestrian accidents: "Pedestrian Crash Avoidance Mitigation" (PCAM) [GAN 06, HAM 15]. To this end, research into PCAM systems has preferably focused on detecting pedestrians, rather than on predicting their walking direction. Therefore, only a few intend to include pedestrian orientation in these systems. Additionally, existing PCAM systems are designed for well-structured areas containing road signs and markings, but in low and middle-income countries, roads are generally poorly structured, as is the case in Morocco.

In this chapter, Morocco was chosen as a case study. According to the Moroccan Ministry of Environment, Transport, Logistics and Water (METLE), in 2016, 28% of Moroccan fatal road accidents were somehow related to pedestrians [MET 17]. In this context, we have collected a new set of data on Moroccan pedestrians from two cities (Rabat and Kénitra). After analyzing this data, we found that traffic laws are less respected by drivers and pedestrians where pedestrians share the roadway with vehicles, especially in poorly structured areas. Due to this behavior, a high rate of pedestrian accidents is observed in Morocco.

The aim of this work is to protect the pedestrian, the most vulnerable road user, particularly when roadside infrastructure lacks pedestrian safety measures, inducing the pedestrian to cross over in an unsafe manner.

For this, Pedestrian Crash Avoidance Mitigation Systems (PCAM) are used by several high-end automobile manufacturers, with the aim of preventing pedestrian-vehicle accidents, and for reducing the harm caused to the pedestrian.

Despite the fact that ADAS systems have had considerable success in preventing accidents involving autonomous vehicles, it has not been so remarkable in reducing pedestrian-related accidents. In March 2018, an Uber autonomous vehicle killed a pedestrian in the state of Arizona in the United States, thus becoming the first autonomous vehicle to kill a pedestrian. The National Transportation Safety Board investigation summarized in the report revealed that the vehicle failed to properly classify the victim, since she was not walking on a pedestrian crossing, a required feature for classifying pedestrians [NAT 18].

Figure 6.1 is a satellite image taken by Google Earth a few seconds before the accident, where we can see the location of the vehicle and the pedestrian as early as 4.2 seconds before the accident, up until the accident. According to this image, we can also notice that the pedestrian crossed diagonally across the middle of the road.

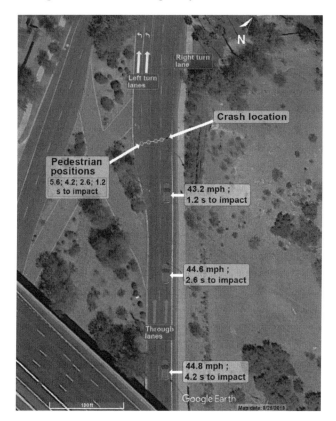

Figure 6.1. *Aerial view showing the location of the accident and the pedestrian's trajectory, the vehicle and the vehicle's speed as early as 5.6 seconds before the accident [NAT 18]*

In this context, we are offering a new PCAM system which takes into account both the presence of unruly pedestrians and of less structured roads.

Our main contributions to this work also include:

– the creation of the first Moroccan pedestrian direction database, called SafeRoad. It was made from scenes collected from different Moroccan cities, using a video recording from an embedded camera inside a moving vehicle;

– and the proposal of a new technique for detecting pedestrian orientation, using the Capsule networks. This technique can be integrated into an ADAS to detect unruly pedestrians and to alert the driver of their presence. The learning and evaluation of this technique is carried out on our SafeRoad database as well as on the Daimler database.

6.2. Advanced Driver Assistance System (ADAS)

ADAS is a system which is directly related to safe driving, improving road safety by warning the driver before a probable collision, aiming to avoid it completely or to reduce damage in case it becomes inevitable. This type of system requires a wide variety of sensors (laser, radar, cameras, etc.) to determine whether the vehicle is in danger of collision with another object. A promising trend in these systems is the creation of systems that analyze driver behavior and automatically adapt it. These systems include:

– *Lane Departure Warning (LDW)*: evaluates whether the vehicle is drifting from the lane, and alerts the driver. The system scans both sides of the vehicle and detects the moment when a vehicle is leaving the lane or the road. By controlling the movement of the steering wheel, the system can tell whether the lane change is intentional or not. In the event that the change is involuntary, the system alerts the driver with a visual or a haptic warning (steering wheel vibrations).

– *Traffic Sign Assist (TSA)*: this application automatically detects traffic signs and can process the information these signs contain, informing the driver about the legal speed limit and priority rules.

– *Blind Spot Detection (BSD)*: monitors the area next to the vehicle. Its function is to warn the driver about an object in the blind spot, using a visual or an audible signal. Its aim is to avoid accidents, especially when changing lanes in heavy traffic conditions.

– *Adaptive Cruise Control (ACC)*: controls the distance in relation to the vehicles ahead, either by alerting the driver or by slowing down the vehicle if the relative distance becomes too small. This mechanism relies on a preset distance parameter, and an alert message is issued if emergency braking is required.

– *Pedestrian Crash Avoidance Mitigation (PCAM)*: pre-collision system enabling the recognition of pedestrians and bicycles in a vehicle's way. It is advisable to alert the driver of a possible head-on collision. If the risk is extremely high, the system can automatically apply the brakes.

Our work focuses primarily on PCAM, while systems currently available focus on pedestrian detection. Recognizing the pedestrian's orientation can be an important asset for reducing pedestrian-vehicle accidents, especially if the pedestrian is unruly and the roads are poorly structured, which explains the genesis of this work.

6.3. Proposal for an applicable system to the Moroccan case

Considered as a "soft user", the pedestrian is the most vulnerable road user. In order to protect the pedestrian, Crash Avoidance Mitigation Systems (PCAM) are used by several high-end automobile manufacturers, thus preventing pedestrian-vehicle accidents and reducing potential harm.

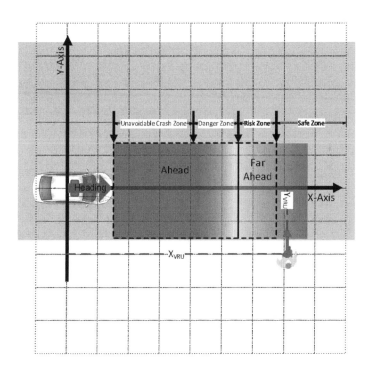

Figure 6.2. *The four danger zones between the vehicle and the pedestrian [TAH 17]*

These systems use visual methods to detect the pedestrian and calculating the distance between them and the vehicle. According to this distance and to the vehicle's speed, four zones can be defined (see Figure 6.2). As specified in [TAH 17], the level of danger for the pedestrian can be determined depending on these factors.

– The first area shown in red (Figure 6.2) is an area where the accident is unavoidable, in other words, if a pedestrian is detected in this area even using the hardest brake possible, the presence of an ADAS cannot be useful in this case. To determine the length of this zone, the vehicle's minimum stop time (TTS$_{min}$) is calculated, which is equal to the time required for the vehicle to stop using the maximum deceleration to brake the vehicle (d$_{max}$) based on current speed (v$_{cur}$), current acceleration (a$_{cur}$), and the driver's reaction time (T$_{DRD}$).

$$TTS_{min} = -\frac{v_{brk}}{d_{max}} + T_{DRD} \qquad [6.1]$$

(v$_{brk}$) is the vehicle's speed after the driver's reaction, which is equal to:

$$v_{brk} = a_{cur} \times T_{DRD} + v_{cur} \qquad [6.2]$$

The distance crossed during this TTS$_{min}$ represents the minimum distance for the vehicle to stop (DTS$_{min}$). Therefore, the length of the zone is defined by

$$DTS_{min} = D_{DRD} + \left(-\frac{v_{brk}^2}{2 \times d_{max}}\right) \qquad [6.3]$$

D$_{DRD}$ is the distance crossed after the driver's reaction:

$$D_{DRD} = \frac{1}{2} \times a_{cur} \times T_{DRD}^2 + v_{cur} \times T_{DRD} \qquad [6.4]$$

– The second zone is the danger zone; in this zone certain accidents can be avoided if the driver is alerted in time. The length of this zone is defined by the distance crossed during the Tdz time, which is equal to the minimum stop time TTS$_{min}$, plus the time T required to alert the alert driver:

$$T_{dz} = TTS_{min} + T_{alert} \qquad [6.5]$$

– The third zone represents the risk zone where danger is not imminent as in the second zone, but which still requires the driver to be alerted in case a pedestrian appears. The length of this zone depends on the time the vehicle can take to brake smoothly d$_{mod}$, which is represented by:

$$T_{risk} = T_{dz} + T_{mod} \qquad [6.6]$$

– The last is a safe zone where no alert is required.

We were inspired by these four severity areas between a vehicle and a pedestrian to design a system which alerts the driver when an unruly pedestrian is present, relying on orientation. An unruly pedestrian crossing between vehicles improperly in the safety zone can increase the risk of an accident. This explains the need to alert the driver about the presence of pedestrians.

A pedestrian can be defined as unruly if:

– They cross the road at an angle which reduces the distance between the pedestrian and the vehicle, increasing the risk of collision. The diagonal crossing is represented by the following pedestrian directions: left-back, right-back, left-ahead, right-ahead.

– When the pedestrian shares the road with other vehicles, the risk of an accident increases by a factor of two, as pointed out by the World Health Organization (WHO) [WOR 13], especially if the pedestrian is walking in the direction of the traffic (represented by [PAI 19] and [SPA 06] as being more risky than walking against the direction of traffic), since vehicles are not visible to the pedestrian in this case. The directions representing a pedestrian walking on the roadway are: back, ahead.

– They cross perpendicularly, away from the passage reserved for pedestrians. This crossing can also be critical if the pedestrian is heading in the direction of the vehicle. Directions are represented by: right, left.

In these cases, the risk of collision depends on the direction of the pedestrian in relation to that of the vehicle, something which explains the benefit of our system.

The pedestrian's direction is recognized by using the Capsule networks.

Within the context of the "SafeRoad" project which is meta-platform for road safety, a study on the behavior of Moroccan pedestrians was conducted, by collecting a database of 5,160 pedestrian images from two Moroccan cities (Rabat and Kénitra), using an industrial camera with a CMOS sensor embedded into a mobile vehicle (Figure 6.3).

The collection was made over three hours of photo-shooting, under different lighting conditions. The choice of the two cities was based on geographical location, since Rabat is the capital city of Morocco. It contains large, keep with well-structured roads, with pedestrian markings on the ground as well as signs. However, during our photo shoot, we captured images of cases where pedestrians were crossing in an unruly manner, putting themselves in danger between vehicles in the middle of the road, despite the presence of a pedestrian crossing. For example,

Figure 6.4 illustrates the case of a woman trying to hail a taxi in the middle of the road. She crosses in the middle of the road between the vehicles in order to approach the taxi, stops between vehicles to make her way to the taxi, and then returns to her initial location, in the same manner.

Figure 6.3. *Image of the camera used inside the vehicle to compile a database of Moroccan pedestrians*

Figure 6.4. *Image showing a pedestrian crossing into the middle of the road diagonally*

This is not the only case, and several pedestrians in the Moroccan capital – which is one of the most sophisticated cities in the country – cross improperly, as shown in Figure 6.5.

Figure 6.5. *Image illustrating unruly pedestrians in the city of Rabat*

The second part of our photo shoot was carried out in the city of Kenitra. This is a city in the region of Rabat, with an area of 672 km^2, 49 km from the capital. However, during the photo shoot at Kénitra we came across places where the roads are poorly structured, containing neither pedestrian crossings nor signals, as seen in Figure 6.6.

Figure 6.6. *Example of a poorly structured road in Kénitra, without markings or signals*

Furthermore, pedestrians cross on the roadway, sometimes leaving the sidewalks empty, without taking the presence of vehicles into consideration. Figure 6.7 illustrates examples of this that came up during the photo shoot.

Figure 6.7. *Images of pedestrian behavior at the city of Kénitra*

Our system detects a pedestrian located in a danger zone, checking whether the pedestrian is heading towards the vehicle's path, and then calculates the time to collision between the vehicle and the pedestrian, as well as the probability of a collision between the two road users. Based on the result of this probability, a decision will be made whether to alert the driver or not.

For this, it is necessary to first detect the pedestrian, and to determine their orientation. This chapter will focus on describing how this is carried out.

As mentioned earlier, a new 100% Moroccan pedestrian database was compiled. These images were extracted from three hours of one-minute video recordings, using an industrial camera containing a CMOS sensor and 2.3 MP resolution and a frequency of up to 60 images per second. However, to ensure high quality, we chose to limit the speed to 30 images per second. Subsequently, the You Only Look Once (YOLO) algorithm [RED 16] was used for detecting and extracting pedestrians. The database totals 5,160 images of pedestrians, resized to 48x48 pixels. The basic idea is to design a system for classifying pedestrians depending on the direction they are moving in. For this, four classes of orientation were defined (right, left, ahead, back), as shown in Figure 6.8.

Figure 6.8. *The four pedestrian directions used*

For classification operations, neural networks have had great success in recent years, especially at the level of "Convolutional Neural Networks" (CNN). The latter have experienced wide applications in image and video recognition. However, convolutional neural networks (CNN) contain a layer called "Pooling" for reducing the network's size. However, this layer generates a loss of information as to the position of objects and their orientation on the input image. As a solution, Capsule networks were proposed by Sara Sabour and Geoffrey Hinton in their article [SAB 17]. A capsule is a group of neurons whose activity vectors represent the entity's position parameters, with the length of these vectors representing the probability of the entity's existence on the input image. Unlike convolutional networks, capsules preserve the detail of the information concerning localization and the entity's position; in other words, a slight rotation of the image implies a slight change on the activation vector. Due to the aforementioned reasons, we chose to use capsule networks for the classification of pedestrian orientation.

The capsule networks represented in [SAB 17] contain an encoder part and a decoder part. The encoder part contains two convolutional layers and a fully connected layer, the main role of this part being classification, whereas the decoder part contains three layers which are fully connected and which are used for reconstructing the input image in order to verify that the right characteristics have been learned.

The first convolutional layer belonging to the encoder part makes it possible to extract the characteristic map from the input image. Then, the result acts as input for the second convolutional layer, called capsule primary layer. This layer contains 32 capsules of 8 dimensions; the vectors of this layer must have a value between 0 and 1 since they represent the entity's probability of existence. For this, a function called squash is applied so as to normalize the vectors between 0 and 1. While the primary role of the third layer is classification, this layer contains one capsule per class, where each capsule contains a membership probability for all classes and the class with the highest probability is assigned to the capsules.

The capsules of the second layer (capsules primary layer) predict the output vectors "$\hat{u}_{j/i}$" of the third layer using their own output vector "u_i" multiplied by the transformation matrix "W_{ij}":

$$\hat{u}_{j/i} = W_{ij} u_i \qquad\qquad [6.7]$$

The transformation matrix "W_{ij}" is learned by the network gradually, using back-propagation during the learning process of the primary capsule layer. As a result of an agreement, "a_{ij}" enters the predicted value "$\hat{u}_{j/i}$" via capsule i at the second layer, and the real value via capsule j at the third layer "v_j", calculated using the following scalar product:

$$a_{ij} = \hat{u}_{ij} . v_j \qquad\qquad [6.8]$$

For each predicted vector, a routing weight called "b_{ij}" is used. It is reset to zero for all capsules on both layers. Then, a softmax function "c_{ij}" is applied to this routing weight for each capsule in the second layer. The weighted sum "s_j" of all the prediction vectors is then calculated for each capsule belonging to the third layer:

$$s_j = \Sigma_i c_{ij} \hat{u}_{j/i} \qquad\qquad [6.9]$$

Then the squash function is applied to this weighted sum, which results in the output vector v_j third layer capsules.

$$v_j = \frac{||s_j||^2}{1+||s_j||^2} \frac{s_j}{||s_j||} \qquad\qquad [6.10]$$

And finally, the routing weight b_{ij} is updated by adding the agreement between the real vector and the reformulated vector p:

$$b_{ij} = b_{ij} + a_{ij} \qquad\qquad [6.11]$$

This whole process represents an iteration of the routing algorithm. In the case of a correct prediction, the routing weight b_{ij} increases, thus also increasing the length of the output vector for the next iteration as well as the entity's probability of existence, represented by such vector.

Then, the length of the vector is used for calculating the probability of the entity's existence, by calculating the loss of margin L_k. A margin loss is calculated for each class k:

$$L_k = T_k max(0, m^+ - ||v_k||^2)^2 + \lambda(1 - T_k)max(0, ||v_k|| - m^-)^2 \quad [6.12]$$

Where:

– $T_k = 1$ if the entity is present;

– m^+ and M^- are hyper parameters which are respectively equal to 0.9 and 0.1;

– $\lambda = 0.5$.

Our pedestrian orientation classification system [DAF 19] based on Capsule networks (CapsNet) contains four layers on the encoder part, as shown in Figure 6.9.

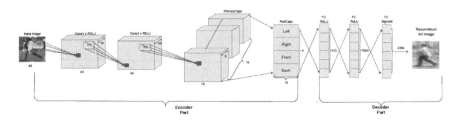

Figure 6.9. *CapsNet architecture used for detecting pedestrian orientation*

The first two layers are convolutional layers with 64x5x5 and 128x5x5 filters, respectively. The first layer takes a gray image of 48x48 as input, extracts the characteristics of the image and sends them as input for the second layer.

The third layer is a primary layer of the capsules containing 16 capsules of 8 dimensions. Each capsule belonging to this layer receives as input the characteristics extracted from the first two convolutional layers.

The fourth layer is called "PedCaps" and contains 4 capsules of 16 dimensions, each capsule referring to a class (four classes of orientations). This layer is used for classifying the input image into one of the four classes mentioned.

The decoder part for reconstructing the input image contains three fully connected layers of 512, 1,024 and 2,304 filters, respectively. The reconstruction of the image is done using the real labels from the PedCaps layer.

To test the approach we used our own database, as well as the public Daimler database for evaluation purposes [ENZ 09]. The network was tested using several capsule networks architectures presented in Table 6.1. According to the results shown in Table 6.1, reducing the number of feature maps from 256 to 128, as well as the number of capsules from 32 to 16 using two iterations of routing, increases accuracy from 92.62% to 96.87%. The architecture used for this system is the third architecture containing two convolutional layers with 64 filters and 128 filters respectively, and 16 primary capsules on the primary layer of the capsules, with an accuracy of 97.60% for the Daimler database.

Architecture	Conv layer No.	Filter No.	Primary Capsule No.	Loss	Accuracy
A1	1	256	32	0.07	90.62%
A2	1	128	16	0.016	96.87%
A3	1	64	8	0.02	96.66%
A4	2	Conv1: 256 and Conv2: 128	16	0.06	95.20%
A5	2	Conv1: 64 and Conv2: 128	16	0.014	97.60%

Table 6.1. *Details of the pedestrian orientation classification based on different architectures*

We compared the proposed approach to some CNN architectures widely used for classifying images. Table 6.2 shows the results obtained for the AlexNet and ResNet architectures on the Daimler database with an accuracy of 95.52% and 96.45% respectively, whereas the accuracy obtained by the Capsule networks exceeded these two CNN architectures. The model is evaluated by calculating the percentage of true positives, which represents the above-mentioned accuracy rate. The ground truth that we have is in the form of images of pedestrians, tagged according to their orientation. If the prediction of the model on a pedestrian image is suitable for the orientation that we have in the ground truth, it will constitute a true positive. Tables 6.3, 6.4 and 6.5 represent the confusion matrices of the three architectures on the Daimler database.

Architecture	Accuracy
AlexNet	95.52%
ResNet	96.45%
CapsNet	97.60%

Table 6.2. *Comparison of the accuracy of AlexNet, ResNet and CapsNet architectures on the Daimler database*

	Ahead	Back	Left	Right
Ahead	1	0	0	0
Back	0.008	0.983	0	0.008
Left	0.020	0.020	0.950	0.008
Right	0.016	0.045	0.012	0.925

Table 6.3. *CapsNet Architecture Confusion Matrix on the Daimler database*

	Ahead	Back	Left	Right
Ahead	0.99	0	0	0.008
Back	0.13	0.97	0.008	0.008
Left	0.016	0.012	0.95	0.02
Right	0.0025	0.05	0.029	0.88

Table 6.4. *AlexNet Architecture Confusion Matrix on the Daimler database*

	Ahead	Back	Left	Right
Ahead	1	0	0	0
Back	0.004	0.98	0	0.008
Left	0.016	0.004	0.97	0
Right	0.008	0.04	0	0.93

Table 6.5. *Resnet Architecture Confusion Matrix on the Daimler database*

	Ahead	Back	Left	Right
Ahead	0.758	0.158	0.062	0.020
Back	0.095	0.779	0.058	0.066
Left	0.116	0.120	0.683	0.079
Right	0.070	0.125	0.083	0.720

Table 6.6. *CapsNet Architecture Confusion Matrix on the Moroccan database*

By comparing the results of the Moroccan database with the Daimler database, we found an accuracy of 73.64% and 97.60% on the two databases respectively. Table 6.6 shows the CapsNet architecture confusion matrix on our collected base.

This difference in the result between the two bases can be interpreted by the difference in the way pedestrians cross on the two bases. The images in the Daimler database show pedestrians generally crossing horizontally, so that the direction is very clear. On the other hand, the images in our database represent pedestrians crossing obliquely, thus producing a bad classification on the four chosen classes of orientation. These bad classifications can be especially noticed between two adjacent directions. Figure 6.10 illustrates examples of misclassification where one backwards direction was classified as being right, and two left directions shown as being ahead (ahead).

Figure 6.10. *Examples of images representing a misclassification*

This proves that the four defined pedestrian directions are not respected in Morocco, since pedestrians cross diagonally, for which the use of eight directions is highly recommended.

6.4. General conclusion

In this chapter we have focused on pedestrian-vehicle collisions, unluckily very prevalent on less structured roads, which often lack measures to ensure the safety of pedestrians.

Our first major contribution is the creation of the first Moroccan database for pedestrian directions, called SafeRoad. This database was collected from several Moroccan cities using an embedded camera inside a moving vehicle. This database was used for learning and testing the system for detecting the orientation of pedestrians using the Capsule networks. This is our second contribution.

The approach aimed to classify the direction of a pedestrian into four orientations (ahead, right, left, back). Capsule networks obtained the best results in terms of accuracy, with 97.60% on the Daimler base and 73.64% on our SafeRoad database. This can be compared to 95.52% and 96.45% respectively, for the AlexNet and ResNet networks on the Daimler database. The results show that the SafeRoad database has a lower accuracy compared to that of Daimler. This difference in results may be due to the different environments in which both databases were collected.

For future research, we intend to extend the detection of pedestrian orientation captured in images to video sequences, in order to be able to integrate it into an ADAS system and implement it in one of the autonomous driving research simulators.

6.5. References

[DAF 19] DAFRALLAH S., AMINE A., MOUSSET S. *et al.*, "Will Capsule Networks overcome Convolutional Neural Networks on Pedestrian Walking Direction?" *International Conference of Intelligent Transportation Systems (ITSC)*, October 2019.

[ENZ 09] ENZWEILER M., GAVRILA D.M. "Monocular pedestrian detection: Survey and experiments", *IEEE Transactions on Pattern Analysis and Machine Intelligence*, vol. 31, no. 12, pp. 2179–2195, 2009.

[GAN 06] GANDHI T., TRIVEDI M.M., "Pedestrian collision avoidance systems: A survey of computer vision based recent studies", *Intelligent Transportation Systems Conference, ITSC'06 IEEE*, Toronto, Ontario, Canada, pp. 976–981, 2006.

[HAM 15] HAMDANE H., SERRE T., MASSON C. *et al.*, "Issues and challenges for pedestrian active safety systems based on real world accidents", *Accident Analysis & Prevention*, no. 82, pp. 53–60, 2015.

[MET 17] METLE, *Recueil des statistiques des accidents corporels de la circulation de l'année 2016*. Ministry of Environment, Transport, Logistics and Water, 2017.

[MOU 12] MOUJAHID M., "Accident : 4 222 tués et 12 500 handicapés à vie par an. Jusqu'à quand", *La vie éco*, 2012.

[NAT 18] NATIONAL TRANSPORTATION SAFETY BOARD, National Transportation Safety Board "Vehicle Automation Report", March 2018.

[PAI 19] PAI C.W., CHEN P.L., MA S.T. *et al.*, "Walking against or with traffic? Evaluating pedestrian fatalities and head injuries in Taiwan", *BMC Public Health*, vol. 19, no. 1, pp. 12–80, 2019.

[RED 16] REDMON J., DIVVALA S., GIRSHICK R. *et al.*, "You only look once: Unified, real-time object detection," *Proceedings of the IEEE Conference on Computer Vision and Pattern Recognition*, pp. 779–788, 2016.

[SAB 17] SABOUR S., FROSST N., HINTON G.E., "Dynamic routing between capsules," *Advances in Neural Information Processing Systems*, pp. 3856–3866, 2017.

[SPA 06] SPAINHOUR L.K., WOOTTON I.A., SOBANJO J.O. *et al.*, "Causative factors and trends in Florida pedestrian crashes", *Transportation Research Record*, vol. 1982, no. 1, pp. 90–98, 2006.

[TAH 17] TAHMASBI-SARVESTANI A., MAHJOUB H.N., FALLAH Y.P. *et al.*, "Implementation and evaluation of a cooperative vehicle-to-pedestrian safety application", *IEEE Intelligent Transportation Systems Magazine*, vol. 9, no. 4, pp. 62–75, 2017.

[WOR 13] WORLD HEALTH ORGANIZATION. *W. H. Organisation, Pedestrian Safety: A Road Safety Manual For Decision-Makers and Practitioners*, 3rd edition, Geneva, Switzerland, 2013.

7

Autonomous Vehicle: What Legal Issues?

7.1. Introduction

The debate on the regulation of the so-called "autonomous" vehicle is raging and is a hot topic for both the general public and professionals in the sector, such as car manufacturers and innovative companies. The legal impacts induced by the use of autonomous vehicles on roads which are open to traffic are significant.

The autonomous or connected vehicle raises a certain number of legal issues and questions, particularly relating to:

– the types of vehicles concerned;

– the driving types targeted;

– the notion of the "driver", taking into account various types of participants: "operators", "supervisors", "users";

– the implications of the legal regime in terms of liability and victim compensation;

– new key components and their role;

– types of insurance coverage;

– personal data protection.

It becomes essential to reflect on the evolution of the existing French legal regime before considering the usefulness and need for a modality specifically applicable to vehicle mobility.

Chapter written by Axelle OFFROY.

In a similar way to technologies such as the Internet or, more recently blockchain, the opportunity arises for a legal regime dedicated to new technologies. For some, excessive regulation poses a risk to innovation and raises fears about the possibility of ideas being crushed, whereas others, especially in France, criticize public authorities for misinterpreting what is at stake, when they refuse to endow technology with a comprehensive legal framework.

Beyond these questions, this major innovation raises many safety-related fears: not only in terms of human/technology interactions, but also in terms of the technology itself. In a global context where cybercrime and hacking are on the rise, it becomes crucial for key components to guarantee the flawless safety of autonomous vehicles.

Finally, technology and the law should not ignore ethics-related issues.

How should we grasp and understand all of the challenges posed by artificial intelligence in the event of an accident? Would it be desirable to allow the artificial intelligence of the autonomous vehicle to protect one individual, rather than another?

In this regard, one of the teams of international researchers from the eminent Massachusetts Institute of Technology (MIT) developed an online test in 2016, called the "Moral Machine", which collected the ethical preferences and choices of nearly two million people around the world (233 countries and territories concerned). Individuals were encouraged to make a life and death choice in the event of an accident. Two years later the results were the subject of an astonishing analysis, published in the journal *Nature*. The article revealed how ethics diverge depending on the cultural, economic and geographic background of the interviewed individual. The questions that the Internet users taking part in the questionnaire had to answer are currently totally foreign to human drivers.

Although some may consider the study unrealistic, one question remains: are the people and political powers prepared to address and decide upon such questions?

These and many other questions need to be addressed in order to prepare and develop the legal regimes applicable to current and future autonomous vehicles.

7.2. The definition of the so-called "autonomous" vehicle

There is no universal definition of the concept of an "autonomous vehicle".

Autonomy refers to the ability of an object or an individual to govern itself or according to its own rules. In the etymological sense, the term "autonomy" is of Greek origin: *autos* means oneself and *nomos* means the law, the rules.

Therefore, the autonomous vehicle is a vehicle capable of reacting by itself, in accordance with specific rules. According to the rules of civil robotics established by the European Parliament, the autonomy of a robot can be defined as the ability to make decisions and put them into practice in the outside world, independent of any outside control or influence.

Given the transversality of the subject, the definition is actually complex. It is essential to approach such a definition by taking into account technical points on one hand, and legal points on the other.

In practice, the autonomous vehicle refers to several techniques, which is why it has been the subject of a classification designating vehicles according to their level of autonomy.

The international organization called SAE (formerly, Society of Automotive Engineers, headquartered in the United States) has developed a classification grid listing five vehicle categories, thus establishing different levels of autonomy.

The last level contemplates the possibility of the vehicle's driver completely delegating the driving task to the vehicle. The latter, or rather the system governing it, can take all of the emergency measures required to preserve the safety of the vehicle and its passengers.

The prospect of a level five autonomous vehicle (full automation) fuels the wildest expectations. However, some believe that such a vehicle will never come to exist, considering that its ability to adapt to external constraints poses immense challenges. This is the case for Luc Julia, one of the inventors of the voice assistant Siri and current vice president of innovation at Samsung, who explains in an extremely pragmatic way that a vehicle with such a level of autonomy cannot see the light of day, given the constraints related to both road infrastructure and individual habits. In particular, he referred to the technical difficulties posed to the system when the vehicle has to move around a roundabout.

From a legal point of view, the expression "autonomous vehicle" may appear imperfect. The term "control delegation" would seem more appropriate. It is this term that has been broadly retained and consecrated by the French legislation in various texts.

7.3. Legal framework and experiments

The driving conditions of autonomous vehicles can currently be found in several texts:

– Law No. 2015-92 of August 17, 2015 relating to energy transition (updated in 2018);

– Ordinance No. 2016-1057 of August 3, 2016 relating to the testing of vehicles with control delegation on public roads;

– Ordinance No. 2018-211 of March 28, 2018 relating to the testing of vehicles with control delegation on public roads;

– Decree of April 17, 2018 relating to the experimentation of vehicles with control delegation on public roads;

– PACTE law of May 22, 2019 relating to the growth and transformation of companies, enumerates the conditions for issuing the authorization intended to ensure the safety of the testing phase.

These texts authorize the public road circulation of vehicles with total or partial control delegation, for testing purposes, that is to say, for the purpose of carrying out technical tests or evaluating the performance of these vehicles.

The test is subject to prior authorization, issued by the Minister responsible for transportation, as well as by various authorities (road manager, competent authority in matters of traffic police, etc.)

The PACTE law specifies the traffic conditions without disrupting the existing laws. It lays down the principle of authorization for autonomous vehicles (private and public transport) to travel on public roads, subject to obtaining a special authorization.

The legal framework imposes a strict condition: the control delegation system must have the ability to be neutralized or deactivated at any moment by the driver.

In the event that the driver is outside of the autonomous vehicle, taking back its control is mandatory.

Finally, the PACTE Law introduces a major provision. It relieves the driver of an autonomous vehicle of his criminal responsibility, when the control delegation system is activated. Criminal responsibility is then transferred to the holder of the authorization for testing purposes.

7.4. The notion of the "driver"

The notion of the "driver" is a major one in the law of civil liability. It is indeed around this key component that road regulations were built.

The driver is commonly defined as *"the natural person who takes the seat in front of the steering wheel and handles the vehicle's control and steering devices"* (Automobile Insurance Manual, 4th J. Landel and L. Namin Edition: Argus Insurance).

The Vienna Convention, adopted on November 8, 1968, aims to facilitate international road traffic and increase road safety through the adoption of uniform rules between the contracting parties.

The Vienna Convention defines the driver as *"any person who drives a motor vehicle (including a cycle), or who guides cattle, singly or in herds, or flocks, or draught, pack or saddle animals on a road. They must remain in control of their vehicle and shall possess the necessary physical and mental ability and be in a fit physical and mental condition to drive."*

However, a question arises in relation to the autonomous vehicle: can the individual behind the wheel be considered a driver? With the control delegation or autonomous car, a shift takes place. The maneuvers originally operated by the driver of the vehicle are/will be transferred, completely or partially, to the intelligent system, depending on its automation level.

On March 23, 2016, the United Nations Economic Commission for Europe adopted a draft revision of the Vienna Convention authorizing autonomous driving. It is still up to the member states to ratify this revision.

Article 125 of the PACTE law of 2019 conditions the issuance of the authorization for the autonomous vehicles' entry into service, with the principle that *"the control delegation system can be neutralized or deactivated at any time by the driver"*. However, the law specifies that *"in the absence of a driver on board, the applicant provides the evidence to certify that a driver [is] located outside the vehicle."*

The notion of human "supervisor" then takes on its full meaning. If the driver is outside of the vehicle, they (or another individual) should supervise the vehicle during the test and be able to regain control at any time and intervene immediately in the event of a failure.

The location of the driver, in the physical sense of the term, is therefore immaterial.

However, the requirement of a driver who is a natural person seems to be maintained in the current state of the texts, the driver's quality is not determined for legal persons.

7.5. The notion of the "custodian"

It is common to assimilate the notion of the custodian to that of the owner, and this follows certain logic. The original concept of custody is based on the responsibility of the owner of the thing. However, this notion has shown certain limitations, in the sense that the owner was held accountable, even when they were not driving the vehicle.

The concept has evolved. Legal custody is characterized by the powers of use, control and direction that a person exercises over a thing or a person. This is the definition adopted by the *Cour de Cassation* (CCass judgment, 2nd Civil Chamber, October 19th, 2006, Appeal No. 04-141777).

In terms of autonomous vehicles, the whole, or part of the driving task can be transferred. Alongside the driver/passenger and vehicle owner, there are manufacturers, designers, supervisors, software designers, etc.

The distinction made by Goldman's thesis back in 1946 takes on its full meaning. In view of the need to provide for the sharing of responsibility between the different key components, it is important to make a distinction between the custody of the thing (of its structure) and the custody of its behavior.

7.6. What liability regime?

There is no textual indication on civil liability in the context of control delegation cars. This legal vacuum around the concept of liability for autonomous vehicles prompts us to make assumptions, given the existing laws and regimes, namely:

– *Liability for facts*, established by the provisions of Article 1242 from the Civil Code: "*We are responsible not only for the damage we cause by our own actions, but also for that which is caused by the actions of people we are accountable for, or the things under our care.*"

The question comes down to determining the autonomous vehicle's custody. It appears essential to establish a difference between behavior custody (driver/supervisor/user) and structure custody (manufacturer/car designer/programmer and subcontractors).

This distinction will be necessary for determining and identifying cascading responsibilities.

– *Liability for defective products*, established by the provisions of Article 1245 from the Civil Code: "*The producer is liable for any damage caused by a defect in his product, whether or not he is bound by a contract with the victim*" and Article 1386-1 from the Civil Code: "*The producer is liable for damage caused by a defect in his product, whether or not he is bound by a contract with the victim.*" This regime provides for a manufacturer liability presumption for any damage caused by a defect in the product. The autonomous car can be identified as a "product" within the context of the aforementioned liability definition. Nonetheless, evidence for the three conditions mentioned is required: the existence of a material fault related to a defect in the product's safety, proof of damage and the causal link between the two.

– *Badinter Law No. 85-677 of July 5, 1985* aimed at improving the situation of traffic accident victims and accelerating the compensation procedure, which applies in the event of traffic accidents involving a land motor vehicle (LMV). The law establishes a special compensation scheme for traffic accident victims. This regime makes it possible to repair the damage and consequences of a traffic accident. Three conditions must be fulfilled: the involvement of a LMV, an accident and the involvement of the LMV in the accident itself. However, legislation does not seem to be strictly adapted to the case of autonomous vehicles. The distinction made between the driver and non-driver should be abolished in order to provide for a more uniform compensation scheme reflecting different cases, integrating the concept of "passenger", in particular.

– *Liability for the actions of a "robot"*. The use of autonomous vehicles and of new technologies in general, has given rise to the idea of a special regime, contemplating the robot's own responsibility, endowed with a legal personality.

This reflection initiated at the European level reaches far beyond the strict framework of autonomous vehicles and traffic accidents. It focuses on robotics and on Artificial Intelligence in the broader sense.

This idea was relayed on the adoption of a resolution by the Parliament, on February 16, 2017[1], providing recommendations to the European Commission as to the civil law rules applicable to robotization. The robot could be placed on a dedicated register and endowed with a capital, just like the current legal persons. It could also benefit from its own insurance.

This decision was heavily criticized. In a letter sent on April 14, 2018 to the European Commission, more than 200 artificial intelligence experts, lawyers, industrialists, scientists and philosophers from 14 countries warned about the risk of endowing robots with a legal status. In addition to the questions relating to ethics, as well as those concerning the rights and duties that should accompany such a regime, there is the question of the ambiguity of such a status.

The European Commission's draft Artificial Intelligence, made public on January 17, 2020, adopts the following definition of Artificial Intelligence:

> *software systems (possibly hardware) designed by human beings, having received a complex goal, deploy actions in the real or digital world by perceiving their environment thanks to data acquisition, by interpreting the structured or unstructured data they collect, by applying reasoning to knowledge or to information processing, inferred from this data and deciding on the best action(s) to be taken in order to achieve the preset goal.*

In its draft, the Commission ultimately excludes attributing a legal personality to the robot, as previously envisaged by Parliament.

7.7. Self-driving vehicle insurance?

According to the specialists, the widespread use of autonomous vehicles should ultimately reduce, or even cause a drop in the number of accidents. The expected effects in terms of road safety have helped to structure their deployment.

According to a study published by KPMG (Autonomous Vehicle Readiness Index, 2nd Edition, 2019), human error is believed to be the cause of nearly 90% of automobile claims.

In relation to road traffic, France has introduced the principle of compulsory insurance, in order to protect the owner of a vehicle against accidents involving it. This compulsory legal framework was introduced by the Law No. 58-208 of

---

1 European Parliament resolution of February 16, 2017, containing recommendations to the Commission regarding civil law rules applicable to robotics (2015/2103 (INL)).

February 27, 1958, and was established in all of the member states of the European Union, via a Directive No. 72/166/EEC on April 24, 1972.

Compulsory insurance, commonly known as "third party insurance", concerns any machine qualified as a land motor vehicle (known as "LMV"), which is governed by the provisions of the Insurance Code and the Highway Code.

The LMV is respectively defined as:

– *"Any self-propelled vehicle intended to travel on the ground, which can be actuated by mechanical force without being linked to a railway track, as well as any trailer, even unhitched"* (Article L211-1, Insurance Code);

– *"Any land vehicle equipped with a propulsion engine, including trolleybuses, traveling on the road by its own means, with the exception of vehicles traveling on rails"* (Article L110-1, Highway Code).

Like any other land motor vehicle, autonomous vehicles must be subject to compulsory insurance. However, in France, there is currently no insurance specifically designed for autonomous vehicles. The use of autonomous vehicles will shake up the world of insurance, and will particularly impact the notion of "risk".

Assuming that the Land Motor Vehicle (LMV) criteria, as determined by the provisions of the Insurance Code and the Highway Code, makes room to consider that the autonomous vehicle meets the definition of a "vehicle" covered by compulsory insurance, the question concerning the legal person subject to such insurance is still open.

Indeed, the French Insurance Code states that the insurance obligation weighs on a natural person, not on the vehicle itself. However, the active role of Artificial Intelligence, resulting in the driver's partial or total passivity, is consequential for the insurance coverage intended to protect the driver of the autonomous vehicle.

The question of determining who is liable for the accident involving autonomous vehicles is also complex, given the multiplicity of stakeholders: operator, supervisor, user, driver, custodian, manufacturer, subcontractors, software publisher, etc.

The implementation of optional insurance alongside compulsory insurance is perfectly possible in regard to autonomous vehicles. Indeed, the principle of contractual freedom remains, subject to compliance with the formal rules prescribed by the law and the Insurance Code.

It will be up to insurers to envision the insurance of the future. The insurance will integrate not only modified risks, but also new ones, such as those related to the connected vehicle and cyber risks. The autonomous vehicle will presumably not escape the curiosity of the hacker community (cybercrime, hacking, vehicle control, theft of personal data, etc.). The automotive sector key components are well aware of safety-related issues. Manufacturers, public authorities, the French supervisory authority, the CNIL, and the National Agency for the Security of Information Systems (ANSSI) are concentrating their efforts on the subject.

7.8. Personal data and the autonomous vehicle

May 25, 2018 is an important date for the protection of personal data. Since that date, the General Data Protection Regulation ("GDPR") of April 26, 2016 has been applicable to the processing of personal data carried out within the framework of the activities of an institution, of a data manager processing information in the territory of the EU, and also outside the EU area.

This European text marks a profound change in terms of formalities. From now on, any data controller referred to in the text must comply with it. The declarative principle which prevailed until then has disappeared.

Article 4 of the GDPR defines personal data quite broadly. Personal data means:

> *any information relating to a natural person, identified or identifiable; data subjects are identifiable if they can be directly or indirectly identified, especially by reference to an identifier such as a name, an identification number, location data, an online identifier or one of several special characteristics, which expresses the physical, physiological, genetic, mental, commercial, cultural or social identity of these natural persons.*

Let us quickly recall that the name, age, email address, IP address, but also driver's license number, geolocalization data, etc. are considered as personal data.

Autonomous vehicles connected and equipped with sensors and a black box are not exempt from regulations on personal data. To be able to function and ensure certain functionalities, be they fundamental or not, data has to be collected and processed.

Among vehicle data there is personal data, insofar as they enable cross-checking to identify an individual, and more precisely the vehicle's owner/driver.

Not only do we refer to the data obtained from the sensors, whose role and functions vary depending on the vehicle's autonomy and make it possible to "track" driver behavior, but also to the data resulting from the vehicle's geolocalization, and the journeys performed.

Therefore, it cannot seriously be refuted that the collection and processing of this data appears to be a major risk of privacy protection breach.

The legal and regulatory framework will be committed to protecting the individual from abusive and indefinite collection. The data retention limit is an essential condition for the viability of the autonomous vehicle. The key components must ensure data safety and offer all means possible to secure and protect privacy.

Data protection, in accordance with the provisions of the GDPR, must be contemplated from the very first moment the designing of the autonomous vehicle begins. The key components must provide useful and necessary functions to ensure the vehicle's compliance with the regulations in force.

There is one equipment device, in particular, which must summon all attention. This is the black box of the autonomous vehicle: the data logger.

Article 11. of a French decree (March 28, 2018) aimed at experimenting with control delegation vehicles on public roads states that "*vehicles should be equipped with a recording device, making it possible to determine, at any time, whether the vehicle has been driven in partial or complete control delegation mode.*"

The data collected by the data logger can be of a different nature. There will be the data generated by the vehicle itself, but there will also be the data generated by other vehicles.

As with sensors, it is an essential tool for locating the sources of autonomous vehicle malfunction. Apart from that, it is of major interest in the reconstruction of traffic accidents.

The data logger, as well as the various loggers already present in the vehicle, helps to precisely identify certain sources of malfunctions and represent a relevant method for gathering evidence.

It is a fundamental tool in the context of automotive expertise, for the analysis of vehicle geolocalization data, the activation/deactivation of the vehicle's automation systems or other related functions, determining the actions of the driver on the vehicle (braking system, use of the telephone), etc. Finally, there is one type of evidence that will enlighten the judge in the case of litigation, and determine (civil and criminal) liabilities in the event of a traffic accident.

The decree of March 2018 states that the data recorded during the five minutes preceding an accident may be kept for one year. Currently, data in the event of an offense is not accessible to insurers without the motorist's consent. Insurers, key components in the victim compensation process, are pushing for access to crucial information. According to them, such devices would have the advantage of making it easier to determine the chain of responsibility in the event of an accident. There is a complex chain of responsibility when it comes to autonomous vehicles, given the multiplicity of players and stakeholders.

However, this device, as virtuous as it is, should not make us forget the consequences it would have in the event of poorly calibrated and ill-controlled use.

As always in matters of freedom, arbitration will take place between the various freedoms involved: respect for private life against freedom of information. The implementation of the protection of the best interests will have to compromise with the part of subjectivity this can induce.

In addition to the needs to limit data collection, ensure the data's integrity and provide for its erasure, it appears essential to maintain limited and controlled access to the data collected by the recorder.

The Mobility Orientation Law (MOL) of December 24, 2019, makes data from devices accessible in the event of an accident. So

> *only the data strictly necessary to determine the activation or not of vehicle control delegation is made accessible, for the purpose of compensating the victims in application of Law No. 85-677 of July 5, 1985 tending to the improvement of the situation of traffic accident victims and the acceleration of compensation procedures.*

This data is accessible to judicial police officers and agents, bodies responsible for investigation and security, insurance companies and guarantee funds for compulsory insurance. The purpose of this data is to determine liabilities.

The MOL introduces measures concerning autonomous vehicles. Thus, under the conditions provided for by the Constitution, the Government is authorized to take any measure falling within the scope of the law, in order to adapt the legislation, and this, within 24 months since the promulgation of such law.

To conclude, the massive introduction of data loggers in autonomous vehicles must be strictly supervised to ensure compliance with the provisions on protection of personal data, and more generally, privacy protection. These limitations, such as to ensure compliance with regulations on the protection of personal data and the respect for privacy, will make this system an essential tool.

7.9. The need for uniform regulation

In its motion for a resolution containing recommendations to the Commission, concerning the civil law rules on robotics of February 16, 2017 (2015/2013 INL), the European Parliament has highlighted the challenges posed by advances in robotics and Artificial Intelligence:

> *Now that humanity finds itself on the dawn of an era where robots, intelligent algorithms, androids and other forms of artificial intelligence, increasingly sophisticated, seem to be about to ignite a new industrial revolution, which will most likely affect all layers of society, it is of fundamental importance for the legislator to examine the consequences and legal and ethical effects of such a revolution, without stifling innovation.*

The issue of the articulation of European and transnational legal regimes cannot be ruled out. In the absence of a uniform legal regime, autonomous vehicles' entry into service in territories, particularly in Europe, could jeopardize the generalization of their use; this is naturally undesirable, as expectations are high.

On the occasion of the World Forum of the United Nations Economic Commission for Europe for the Harmonization of Vehicle Regulations, in June 2020, more than 50 countries, including France, adopted binding regulations on automated lane-keeping systems for "level 3" cars, including mandatory black boxes. The text introduces strong safety requirements, including the speed limit to 60 km/hour. This regulation will come into force at the beginning of 2021. This new international text marks an important step in the deployment of autonomous vehicles.

List of Authors

Aouatif AMINE
LGS
ENSA-Kenitra
Morocco

Thierry BAPIN
NextMove
Normandy
France

Abdelaziz BENSRHAIR
LITIS/ITI/INSA
Rouen
France

Jean-Sébastien BERTHY
ESI Group
Nantes
France

Fabien BONARDI
IBISC Laboratory
Paris-Saclay University
University of Évry
France

Samia BOUCHAFA
IBISC Laboratory
Paris-Saclay University
University of Évry
France

Alberto BROGGI
VisLab
University of Parma
Italy

Safaa DAFRALLAH
LGS
ENSA-Kenitra
Morocco

Philippe DESOUZA
ESI Group
Rungis
France

Thierry ERNST
YoGoKo
Paris and Rennes
France

Dominique GRUYER
PICS-L Laboratory
Department of Components
and Systems
Gustave Eiffel University
Versailles-Satory
France

Hicham HADJ-ABDELKADER
IBISC Laboratory
Paris-Saclay University
University of Évry
France

Mokrane HADJ-BACHIR
ESI group
Nantes
France

Serge LAVERDURE
ESI Group
Nantes
France

Stéphane MOUSSET
LITIS/University of Rouen
Normandy
France

Axelle OFFROY
SHYRKA
Rouen
France

Olivier ORFILA
PICS-L Laboratory
Department of Components
and Systems
Gustave Eiffel University
Versailles-Satory
France

Rémi SAINCT
PICS-L Laboratory
Department of Components
and Systems
Gustave Eiffel University
Versailles-Satory
France

Désiré SIDIBÉ
IBISC Laboratory
Paris-Saclay Univeristy
University of Évry
France

Gérard YAHIAOUI
NEXYAD
Saint-Germain-en-Laye
France

Index

3D immersive application, 172

A

Advanced Driver Assistance System (ADAS), 127, 216, 218
Ant Colony Optimization (ACO), 87, 107, 110, 111
artificial intelligence (*see also* eXplainable Artificial Intelligence (XAI)), 234, 239–241, 245
  methods, 82, 87
autonomous (*see also* levels of autonomy)
  driving, 88, 89, 102, 126, 169, 171, 175,
  vehicle, 25, 27, 71, 128, 145, 153, 169, 170, 233–245

C, D

classification, 235
  centralized, 181, 191, 195, 197, 198, 200–202, 205, 209, 210
  hybrid, 192, 200, 201, 205, 208
  localized, 187–191, 195, 197–201, 204, 205, 207, 209, 210, 212
  networked, 197
  secure, 191, 207

co-pilot, 169–171
  virtual, 88, 91, 101
computer vision, 33
conditions
  adverse, 126
  degraded, 126, 127, 136, 139, 146, 150
connected and cooperative vehicle, 125, 158, 176, 182–184, 212
control delegation, 235–238, 243, 244
Cooperative ITS (C-ITS), 181, 186, 188, 192, 196, 197, 203–207, 209–211
  services, 188, 203–205, 211
  standards, 192, 210, 211
  technologies, 209
cybersecurity, 234, 242
driver behavior, 218, 243

E, F, G

eco-mobility, 171–174
ethics, 234, 240
experiments, 236, 243
eXplainable Artificial Intelligence (XAI) , 9, 22
free space, 46

Other titles from

in

Mechanical Engineering and Solid Mechanics

2021

CHALLAMEL Noël, KAPLUNOV Julius, TAKEWAKI Izuru
Modern Trends in Structural and Solid Mechanics 1: Static and Stability
Modern Trends in Structural and Solid Mechanics 2: Vibrations
Modern Trends in Structural and Solid Mechanics 3: Non-deterministic Mechanics

DAHOO Pierre Richard, POUGNET Philippe, EL HAMI Abdelkhalak
Applications and Metrology at Nanometer Scale 1: Smart Materials, Electromagnetic Waves and Uncertainties (Reliability of Multiphysical Systems Set – Volume 9)
Applications and Metrology at Nanometer Scale 2: Measurement Systems, Quantum Engineering and RBDO Method (Reliability of Multiphysical Systems Set – Volume 10)

LEDOUX Michel, EL HAMI ABDELKHALAK
Heat Transfer 1: Conduction (Mathematical and Mechanical Engineering Set – Volume 9)
Heat Transfer 2: Radiative transfer (Mathematical and Mechanical Engineering Set – Volume 10)

FROSSARD Etienne
Granular Geomaterials Dissipative Mechanics: Theory and Applications in Civil Engineering

RADI Bouchaib, EL HAMI Abdelkhalak
Advanced Numerical Methods with Matlab® 1: Function Approximation and System Resolution
(Mathematical and Mechanical Engineering SET – Volume 6)
Advanced Numerical Methods with Matlab® 2: Resolution of Nonlinear, Differential and Partial Differential Equations
(Mathematical and Mechanical Engineering SET – Volume 7)

SALENÇON Jean
Virtual Work Approach to Mechanical Modeling

2017

BOREL Michel, VÉNIZÉLOS Georges
Movement Equations 2: Mathematical and Methodological Supplements
(Non-deformable Solid Mechanics Set – Volume 2)
Movement Equations 3: Dynamics and Fundamental Principle
(Non-deformable Solid Mechanics Set – Volume 3)

BOUVET Christophe
Mechanics of Aeronautical Solids, Materials and Structures
Mechanics of Aeronautical Composite Materials

BRANCHERIE Delphine, FEISSEL Pierre, BOUVIER Salima, IBRAHIMBEGOVIC Adnan
From Microstructure Investigations to Multiscale Modeling: Bridging the Gap

CHEBEL-MORELLO Brigitte, NICOD Jean-Marc, VARNIER Christophe
From Prognostics and Health Systems Management to Predictive Maintenance 2: Knowledge, Traceability and Decision
(Reliability of Multiphysical Systems Set – Volume 7)

EL HAMI Abdelkhalak, RADI Bouchaib
Dynamics of Large Structures and Inverse Problems

(*Mathematical and Mechanical Engineering Set – Volume 5*)
Fluid-Structure Interactions and Uncertainties: Ansys and Fluent Tools
(*Reliability of Multiphysical Systems Set – Volume 6*)

KHARMANDA Ghias, EL HAMI Abdelkhalak
Biomechanics: Optimization, Uncertainties and Reliability
(*Reliability of Multiphysical Systems Set – Volume 5*)

LEDOUX Michel, EL HAMI Abdelkhalak
Compressible Flow Propulsion and Digital Approaches in Fluid Mechanics
(*Mathematical and Mechanical Engineering Set – Volume 4*)
Fluid Mechanics: Analytical Methods
(*Mathematical and Mechanical Engineering Set – Volume 3*)

MORI Yvon
Mechanical Vibrations: Applications to Equipment

2016

BOREL Michel, VÉNIZÉLOS Georges
Movement Equations 1: Location, Kinematics and Kinetics
(*Non-deformable Solid Mechanics Set – Volume 1*)

BOYARD Nicolas
Heat Transfer in Polymer Composite Materials

CARDON Alain, ITMI Mhamed
New Autonomous Systems
(*Reliability of Multiphysical Systems Set – Volume 1*)

DAHOO Pierre Richard, POUGNET Philippe, EL HAMI Abdelkhalak
Nanometer-scale Defect Detection Using Polarized Light
(*Reliability of Multiphysical Systems Set – Volume 2*)

DE SAXCÉ Géry, VALLÉE Claude
Galilean Mechanics and Thermodynamics of Continua

DORMIEUX Luc, KONDO Djimédo
Micromechanics of Fracture and Damage
(*Micromechanics Set – Volume 1*)

EL HAMI Abdelkhalak, RADI Bouchaib
Stochastic Dynamics of Structures
(Mathematical and Mechanical Engineering Set – Volume 2)

GOURIVEAU Rafael, MEDJAHER Kamal, ZERHOUNI Noureddine
From Prognostics and Health Systems Management to Predictive
Maintenance 1: Monitoring and Prognostics
(Reliability of Multiphysical Systems Set – Volume 4)

KHARMANDA Ghias, EL HAMI Abdelkhalak
Reliability in Biomechanics
(Reliability of Multiphysical Systems Set –Volume 3)

MOLIMARD Jérôme
Experimental Mechanics of Solids and Structures

RADI Bouchaib, EL HAMI Abdelkhalak
Material Forming Processes: Simulation, Drawing, Hydroforming and
Additive Manufacturing
(Mathematical and Mechanical Engineering Set – Volume 1)

2015

KARLIČIĆ Danilo, MURMU Tony, ADHIKARI Sondipon, MCCARTHY Michael
Non-local Structural Mechanics

SAB Karam, LEBÉE Arthur
Homogenization of Heterogeneous Thin and Thick Plates

2014

ATANACKOVIC M. Teodor, PILIPOVIC Stevan, STANKOVIC Bogoljub,
ZORICA Dusan
Fractional Calculus with Applications in Mechanics: Vibrations and
Diffusion Processes
Fractional Calculus with Applications in Mechanics: Wave Propagation,
Impact and Variational Principles

CIBLAC Thierry, MOREL Jean-Claude
Sustainable Masonry: Stability and Behavior of Structures

2012

DAVIM J. Paulo
Mechanical Engineering Education

DUPEUX Michel, BRACCINI Muriel
Mechanics of Solid Interfaces

ELISHAKOFF Isaac *et al.*
*Carbon Nanotubes and Nanosensors: Vibration, Buckling
and Ballistic Impact*

GRÉDIAC Michel, HILD François
Full-Field Measurements and Identification in Solid Mechanics

GROUS Ammar
Fracture Mechanics – 3-volume series
Analysis of Reliability and Quality Control – Volume 1
Applied Reliability – Volume 2
Applied Quality Control – Volume 3

RECHO Naman
Fracture Mechanics and Crack Growth

2011

KRYSINSKI Tomasz, MALBURET François
Mechanical Instability

SOUSTELLE Michel
An Introduction to Chemical Kinetics

2010

BREITKOPF Piotr, FILOMENO COELHO Rajan
Multidisciplinary Design Optimization in Computational Mechanics

DAVIM J. Paulo
Biotribolgy

Printed and bound by CPI Group (UK) Ltd, Croydon, CR0 4YY

28/10/2024

14581337-0001